Judith Schenzel

Mycotoxins

Judith Schenzel

Mycotoxins

Overlooked Aquatic Micropollutants

Südwestdeutscher Verlag für Hochschulschriften

Impressum / Imprint
Bibliografische Information der Deutschen Nationalbibliothek: Die Deutsche Nationalbibliothek verzeichnet diese Publikation in der Deutschen Nationalbibliografie; detaillierte bibliografische Daten sind im Internet über http://dnb.d-nb.de abrufbar.
Alle in diesem Buch genannten Marken und Produktnamen unterliegen warenzeichen-, marken- oder patentrechtlichem Schutz bzw. sind Warenzeichen oder eingetragene Warenzeichen der jeweiligen Inhaber. Die Wiedergabe von Marken, Produktnamen, Gebrauchsnamen, Handelsnamen, Warenbezeichnungen u.s.w. in diesem Werk berechtigt auch ohne besondere Kennzeichnung nicht zu der Annahme, dass solche Namen im Sinne der Warenzeichen- und Markenschutzgesetzgebung als frei zu betrachten wären und daher von jedermann benutzt werden dürften.

Bibliographic information published by the Deutsche Nationalbibliothek: The Deutsche Nationalbibliothek lists this publication in the Deutsche Nationalbibliografie; detailed bibliographic data are available in the Internet at http://dnb.d-nb.de.
Any brand names and product names mentioned in this book are subject to trademark, brand or patent protection and are trademarks or registered trademarks of their respective holders. The use of brand names, product names, common names, trade names, product descriptions etc. even without a particular marking in this works is in no way to be construed to mean that such names may be regarded as unrestricted in respect of trademark and brand protection legislation and could thus be used by anyone.

Coverbild / Cover image: www.ingimage.com

Verlag / Publisher:
Südwestdeutscher Verlag für Hochschulschriften
ist ein Imprint der / is a trademark of
AV Akademikerverlag GmbH & Co. KG
Heinrich-Böcking-Str. 6-8, 66121 Saarbrücken, Deutschland / Germany
Email: info@svh-verlag.de

Herstellung: siehe letzte Seite /
Printed at: see last page
ISBN: 978-3-8381-3621-9

Zugl. / Approved by: ETH Zurich, Diss.-Nr.: 20436, 2012

Mycotoxins
Overlooked Aquatic Micropollutants

presented by
JUDITH SCHENZEL

2013

TABLE OF CONTENTS

SUMMARY

Emerging contaminants, or organic micropollutants, are a serious problem in the aquatic environment, because most of these compounds are ubiquitously present and comprise a high bioaccumulation potential. Besides compounds of anthropogenic origin (e.g., pharmaceuticals, pesticides, steroids or personal care products), also naturally produced micropollutants can be present in the environment. One class of naturally produced micropollutants are mycotoxins. Their mobility was first observed in a laboratory-scale leaching study in the 1980s and demonstrated the potential of mycotoxin emission into the aquatic environment.

Mycotoxins are naturally occurring secondary plant metabolites produced by fungi of different genera, and exhibit a great structural diversity, which results in widely varying chemical and physical properties. These toxic metabolites are commonly found in crops grown and stored for human or animal consumption, as well as in processed food, hence their occurrence in food and feed has been studied extensively. However, the environmental exposure to mycotoxins has been scarcely investigated. Thus far, the identified main input sources of mycotoxins into the aquatic environment include 1) run-off and drainage from fields cultivated with fungi-infected cereals, 2) husbandry animals (i.e., excretion from grazing livestock, run-off from feeding operations, or manure application), and 3) human excretion via sewer systems. Especially, the production and emission of various mycotoxins from infected agricultural plots cropped with small grain cereals, as well as their occurrence in surface waters has hardly been investigated systematically.

The objectives of this doctoral thesis were: 1) to obtain experimentally determined soil organic-carbon water partitioning coefficients for a diverse set of mycotoxins to 2) extend the limited knowledge on environmental fate and behavior of a few compounds to a larger variety of mycotoxins, hence natural organic micropollutants, and 3) to quantify their loads in the respective matrices (drainage- and surface waters, waste water treatment plant (WWTP) effluent), 4) to determine the contribution of different sources (i.e., WWTP effluents, and agricultural areas), and 5) to evaluate their ecotoxicological relevance.

In this doctoral thesis, a broad selection of mycotoxins originating from different fungal species was investigated. Target compounds included aflatoxins (aflatoxins B_1, B_2, G_1, G_2, M_1), ochratoxins (ochratoxin A, B), citrinin, patulin, sterigmatocystin, and sulochrin produced by different *Aspergillus* spp. strains. Furthermore, mycotoxins produced by *Alternaria* spp. (alternariol, alternariol monomethylether, altenuene, tentoxin), ergot alkaloids (ergocornine, ergocryptine), and *Fusarium* spp. toxins (fumonisins, beauvericin, resorcyclic acid lactones, type A and B trichothecenes) were aimed at.

First, appropriate analytical methods had to be developed to quantify mycotoxins in aqueous and solid agro-environmental samples. Aqueous samples were solid phase extracted, whereas solid samples were solid-liquid extracted. Matrix-related effects and losses during sample preparation were compensated either by the addition of isotope-labeled analogues or by matrix-matched calibrations. Mycotoxins were further separated and detected by high performance liquid chromatography coupled to tandem mass spectrometry. The method detection limits for the different aqueous matrices (drainage- and surface water, as well as WWTP effluents) were below 10 ng per liter for 27 compounds. The absolute method recoveries for 13 out of the 33 mycotoxins were higher than 70%, and relative method recoveries for seven compounds ranged from 72 to 103% in waste water treatment plant effluent (at 25 ng/L). The method detection limit for whole wheat plants ranged from 1 to 26 ng/$g_{dryweight(dw)}$. Here, rather high matrix effects and high analyte concentrations made the application of isotope-labeled analogues unaffordable. The total method recoveries for 26 out of 28 compounds ranged between 69 and 122% because of the use of matrix-matched calibrations.

The established analytical methods were further applied in two different studies to investigate the emission of mycotoxins from a) an experimental field via drainage water, and b) via WWTP effluents over approximately two years. Additionally, the presence of mycotoxins in surface water samples was investigated over the same time period to attain seasonal concentration dynamics as part of an immission study.

The occurrence and abundance of mycotoxins produced on an artificially infected test field cropped with winter wheat and their emission into the drainage water was determined by applying the respective, validated analytical methods. Eight mycotoxins were repeatedly detected in whole wheat plant samples of varying subfields, which were infected with four different *Fusarium*-strains. Within all subfields, and over the whole investigation period, deoxynivalenol was the dominant mycotoxin, also in drainage water samples. The maximum concentration of deoxynivalenol in wheat plants and drainage water samples were 133 mg/kg_{dw}, and 1.1 µg/L, respectively. Resultant loads of all detected mycotoxins were determined: infected wheat plants produced 2.3-292 g mycotoxins/ha/y, whereof 0.5-354 mg mycotoxins/ha/y, i.e., 0.002-0.12% were emitted with drainage water. Hence, only a minor fraction of the total mycotoxins initially produced by the infected wheat plants was emitted via drainage water. These emitted fractions depended upon several processes such as climatic conditions, crop variety, possible degradation processes (not investigated in this study), and sorption processes. Therefore, sorption coefficients of several mycotoxins between water and a model sorbent for natural organic matter were experimentally determined. This unique data set indicated that for example the leaching of the hydrophilic compounds deoxynivalenol and nivalenol into drainage water can be explained by their low sorptive affinities. However, under good agricultural practice (including plowing, a proper pest management, crop rotation, as well as the cultivation of less *Fusarium*-susceptible wheat varieties), and without any artificial plant infestation, the mycotoxin concentrations and emission should usually be much lower.

Nevertheless, mycotoxin emission from *Fusarium*-infected fields of the investigated river catchment areas was estimated to range between 0.2 – 970, 0.3 – 1105, 0.06 – 270, and 0.003 – 5.8 g, respectively, over the whole period of investigation from December 2009 to October 2011 for DON, NIV, 3-AcDON, and BEA.

In WWTP effluent samples DON again prevailed with maximum concentrations of up to 73 ng/L over the two other detected type B trichothecenes (NIV and 3-AcDON), and BEA. Additonally, the detected mycotoxin concentrations from the sporadically monitored WWTP effluents were translated into daily emitted loads per capita. Hence, these loads can be extrapolated to the whole investigation period as well, which ranged from 0.05-1.5 kg, 0.01-0.4 kg, 0.02-0.6 kg, and 1.3-36 g for DON, NIV, 3-AcDON, and BEA, respectively.

Mycotoxins were also regularly detected in Swiss surface water samples, although at lower concentrations compared to drainage water and WWTP effluents. Nivalenol prevailed in surface waters with maximum concentrations of up to 24.1 ng/L. The cumulative mycotoxin loads over the whole investigation period ranged from 0.1-10.9 kg, 0.1-13.6 kg, not detected-1.0 kg, not detected-4.4 kg for DON, NIV, 3-AcDON, and BEA, respectively. These loads matched the sum of both estimated loads (*Fusarium*-infected crops, and WWTP) within a factor of two for DON. Hence, these two sources largely explain the loads of mycotoxins quantified in the sampled river water even though husbandry animals as a whole could not be taken into account for the source apportionment.

The here presented results document the important role of both investigated sources (WWTP effluents, and *Fusarium* infected crops) for the occurrence of the most prominent mycotoxins detected in surface waters in Switzerland. The contribution from each source varied according to the specific characteristics of the river catchment area, i.e., the number of WWTPs and inhabitants connected to each river sampling station compared to the size of the agricultural area cropped with winter wheat.

Based on the quantified mycotoxin concentrations in drainage water and WWTP effluent samples, as well as the calculated loads from these sources, the mycotoxin exposure to surface waters is rather low. So far, only a few studies are available dealing with the ecotoxicity of these compounds. Nevertheless, these studies indicate, that the ecotoxicological relevance of mycotoxins in the aquatic environment is rather small due to strong dilution in surface waters and the low exposure.

Zusammenfassung

Neu erfasste chemische Verunreinigungen, oft auch als organische Mikroverunreinigungen bezeichnet, sind ein ernstes Problem für die aquatische Umwelt, da die meisten dieser Verbindungen ubiquitär vorhanden sind und zudem ein hohes Bioakkumulationspotential aufweisen. Neben Verbindungen anthropogenen Ursprungs (zum Beispiel: Arzneimittel, Pestizide, Steroide oder Körperpflegemittel), können in der Umwelt auch natürlich produzierte Mikroverunreinigungen auftreten. Eine Klasse natürlich vorkommender Mikroverunreinigungen sind Mykotoxine. Deren Mobilität wurde erstmals in den 1980er Jahren in einer Labor-Auswaschung-Studie beobachtet und demonstriert die potentielle Mykotoxin-Emission in die Gewässer.

Mykotoxine sind natürlich vorkommende, sekundäre Stoffwechselprodukte von Pilzen verschiedener Gattungen, und weisen eine grosse strukturelle Vielfalt auf, welche in ihren verschiedenen chemischen-physikalisch Eigenschaften zu Tage tritt. Diese toxischen Metaboliten kommen häufig in für menschliche und tierische Nahrung angebautem und gelagertem Getreide, sowie in verarbeiteten Lebens- und Futtermitteln vor, weswegen das Vorkommen in diesen ausführlich untersucht wurde. Bisher kaum untersucht ist jedoch die Umweltexposition gegenüber Mykotoxinen. Die bisher identifizierten Haupteintragspfade von Mykotoxinen in die Oberflächengewässer sind: 1) der direkte Austrag aus landwirtschaftlichen Flächen, welche mit von Pilzen infiziertem Getreide bepflanzt sind, 2) Tierhaltung (d.h. Ausscheidung von Nutztieren, direkter Austrag aus Fütterungs- und Mastanlagen, oder das Ausbringen von Stallmist) und 3) durch menschliche Ausscheidung über die Kanalisation. Insbesondere die Produktion und die Emission von verschiedenen Mykotoxinen aus infizierten landwirtschaftlichen Parzellen, die mit Getreide bepflanzt sind, sowie ihr Vorkommen in Fliessgewässern, wurden bisher kaum systematisch untersucht.

Die Ziele dieser Dissertation waren: 1) experimentell organische Kohlenstoff-Wasser-Verteilungskoeffizienten für eine Reihe von Mykotoxinen zu ermitteln, um 2) die begrenzten Kenntnisse über das Umweltverhalten von einigen wenigen Verbindungen zu einer Vielzahl von Mykotoxinen zu erweitern, 3) ihre Frachten sollten in den entsprechenden Matrices (Drainagewasser und Fliessgewässer, Kläranlagenausfluss) quantifiziert werden, sowie 4) der Beitrag aus den einzelnen Quellen (Kläranlagenabwässer und landwirtschaftlichen Flächen) bestimmt werden, und 5) ihre ökotoxikologische Relevanz bewertet werden.

In dieser Doktorarbeit wurde eine breite Auswahl von Mykotoxinen, die von verschiedenen Pilzarten stammen, untersucht. Zielverbindungen umfassten Aflatoxine (Aflatoxin B_{1}, B_{2}, G_{1}, G_{2}, M_{1}), Ochratoxine (Ochratoxin A, B), Citrinin, Patulin, Sterigmatocystin und Sulochrin, welche von verschiedenen *Aspergillus* spp. Stämmen gebildet werden können. Darüber hinaus wurden ebenfalls Mykotoxine von *Alternaria* spp. (Alternariol, Alternariol monomethylether, Altenuene, Tentoxin), Mutterkorn-Alkaloide (Ergocornin, Ergocryptin) und *Fusarium* Toxine (Fumonisine, Beauvericin, Resorzyklische Saure Laktone, Typ A und B Trichothecene) untersucht.

Für die Quantifizierung der Mykotoxine in wässrigen und festen Proben wurden geeignete Analysemethoden entwickelt. Die Festphasenextraktion wurde für wässrige Proben angewendet, während bei festen Proben eine fest-flüssig Extraktion durchgeführt wurde. Matrixbedingte Effekte und Verluste während der Probenaufarbeitung wurden entweder durch die Zugabe von isotopenmarkierten Standards oder mittels auf die Matrix abgestimmten Kalibrierungen kompensiert. Mykotoxine wurden nachfolgend aufgetrennt und mittels Hochleistungs-Flüssigchromatographie gekoppelt an ein Tandem-Massenspektrometer detektiert. Die Nachweisgrenze für die einzelnen wässrigen Matrices (Drainagewasser, Fliessgewässer sowie Kläranlagenausfluss) liegt für 27 Verbindungen unter 10 ng pro Liter. Die absolute Wiederfindungsrate für 13 der 33 Mykotoxine war grösser als 70%, und die relative Wiederfindungsrate für sieben Verbindungen lag zwischen 72 und 103% in Kläranlageabwasser (bei 25 ng/L). Die Nachweisgrenze für ganze Weizenpflanzen variierte von 1 bis 26 $ng/g_{Trockengewicht (TG)}$. Hohe Matrixeffekte und Analytkonzentrationen machten die Verwendung von isotopenmarkierten Standards für die Quantifizierung von Mykotoxinen in ganzen Weizenpflanzen unerschwinglich. Die absolute Wiederfindungsrate lag für 26 von 28 Verbindungen zwischen 69 und 122%, aufgrund der Verwendung von an die Matrix angepassten Kalibrierungen.

Die etablierten analytischen Methoden wurden nachfolgend in zwei verschiedenen Studien eingesetzt, um die Emission von Mykotoxinen a) über das Sickerwasser aus einem landwirtschaftlichen Testfeld, und b) über den Kläranlagenausfluss zu ermitteln innerhalb eines Untersuchungszeitraumes von annähernd zwei Jahren. Im selben Zeitraum wurde zusätzlich das Auftreten von Mykotoxinen in Oberflächengewässerproben untersucht, wobei die saisonale Konzentrationsdynamik im Rahmen einer Immission-Studie ermittelt wurde.

Das Auftreten und die produzierte Menge von Mykotoxinen auf einem künstlich infizierten Winterweizen-Testfeld und deren Austrag in das Drainagewasser wurden mit der jeweiligen, validierten Analysemethode bestimmt. Acht Mykotoxine wurden wiederholt in Weizenpflanzen von unterschiedlichen Teilfeldern gemessen, welche mit vier verschiedenen *Fusarium*-Stämmen infiziert wurden. Innerhalb aller Teilfelder, und über den gesamten Untersuchungszeitraum hinweg, war Deoxynivalenol das dominierende Mykotoxin, ebenso in den Drainagewasserproben. Die maximalen Konzentration von Deoxynivalenol in Weizenpflanzen und Drainagewasserproben waren: 133 mg/kg_{TG} und 1.1 µg/L. Für alle quantifizierten Mykotoxine konnten folgende Frachten bestimmt werden: infizierte Weizenpflanzen produzierten 2.3 bis 292 g Mykotoxine/ha/Jahr, wovon 0.5 bis 354 mg Mykotoxine/ha/Jahr, dass heisst 0.002 bis 0.12% im Drainagewasser emittierten. Daher wird nur ein geringfügiger Teil der insgesamt in den Weizenpflanzen produzierten Mykotoxinen ins Drainagewasser emittiert. Die Grösse der emittierten Fraktion wird von mehreren Faktoren wie klimatischen Bedingungen, Pflanzensorte, möglicher Abbau-Prozesse (in dieser Studie nicht untersucht), und Sorptionsprozessen beeinflusst. Sorptionskoeffizienten zwischen Wasser und einem Modell-Sorbenten für natürliches, organisches Material wurden für mehrere Mykotoxine experimentell bestimmt. Dieses einzigartige Datenset veranschaulicht, dass die Auswaschung von hydrophilen Verbindungen, wie Deoxynivalenol und Nivalenol, in das Drainagewasser auf die geringe sorptive Affinität dieser zurück geführt werden kann. Unter guter landwirtschaftlicher Praxis (einschließlich Pflügen, einer angemessene Schädlingsbekämpfung, Fruchtfolge, sowie dem Anbau von weniger *Fusarien*-anfälligen Weizensorten) und ohne das künstliche Infizieren der Pflanzen, sollten die Mykotoxin-Konzentrationen und die Emission jedoch wesentlich niedriger sein.

Dennoch, die für den gesamten Untersuchungszeitraum von Dezember 2009 bis einschliesslich Oktober 2011 abgeschätzten Mykotoxin-Emissionen von *Fusarium*-infizierten Anbauflächen aus den Einzugsgebieten der untersuchten Fliessgewässer lagen zwischen 0.2 - 970, 0.3 - 1105, 0.06 - 270, und 0.003 - 5.8 g für Deoxynivalenol, Nivalenol, 3-Acetyl-Deoxynivalenol, und Beauvericin.

In Kläranlagenausflussproben dominierte ebenfalls Deoxynivalenol mit maximalen Konzentrationen von bis zu 73 ng/L über die anderen beiden Typ B Trichothecene (Nivalenol und 3-Acetyl-Deoxynivalenol), und Beauvericin. Ferner wurden die gemessenen Mykotoxin-Konzentrationen vom sporadisch beprobten Kläranlagenausfluss in täglich emittierte Frachten pro Kopf umgerechnet. Demzufolge können diese Frachten ebenfalls auf den gesamten Untersuchungszeitraum extrapoliert werden und variieren von 0.05 - 1.5 kg, 0.01 - 0.4 kg, 0.02 - 0.6 kg und 1.3 - 36 g für Deoxynivalenol, Nivalenol, 3-Acetyl-Deoxynivalenol, und Beauvericin.

Mykotoxine wurden auch regelmäßig in Schweizer Oberflächengewässern nachgewiesen, obwohl die Konzentrationen, verglichen mit den Drainagewasser- und Kläranlagenproben, deutlich niedriger waren. Nivalenol dominierte in den Fliessgewässerproben mit maximalen Konzentrationen von bis zu 24.1 ng/L. Die kumulierten Mykotoxin-Frachten über den gesamten Untersuchungszeitraum reichten von 0.1-10.9 kg, 0.1-13.6 kg, nicht detektiert-1.0 kg, nicht detektiert -4.4 kg für Deoxynivalenol, Nivalenol, 3-Acetyl-Deoxynivalenol und Beauvericin. Diese Frachten stimmen für Deoxynivalenol bis auf einen Faktor zwei mit der

Summe der beiden abgeschätzten Frachten (*Fusarium*-infizierten Anbauflächen und Kläranlagenausfluss) überein. Folglich erklären beide Quellen weitgehend die in den Fliessgewässern quantifizierten Mykotoxin-Frachten, obwohl Tierhaltung als Ganzes bei der Quellenzuordnung nicht berücksichtigt werden konnte.

Die hier präsentierten Ergebnisse dokumentieren die wichtige Rolle der beiden untersuchten Quellen (Kläranlagenausfluss und *Fusarium*-infizierten Getreidearten) für das Auftreten der prominentesten Mykotoxine in Schweizer Oberflächengewässern. Der Beitrag der beiden Quellen variiert je nach den spezifischen Eigenschaften des Einzugsgebietes, das heisst der Anzahl von Kläranlagen und Einwohnern die an das Fliessgewässer angegliedert sind, verglichen zur Größe der landwirtschaftlichen Anbaufläche die mit Winterweizen bepflanzt ist.

Basierend auf den quantifizierten Mykotoxinkonzentrationen im Sickerwasser und in den Kläranlagenausflussproben, sowie den berechneten Frachten aus diesen Quellen ist die Mykotoxin-Belastung von Oberflächengewässern eher gering. Bislang gibt es nur wenige Studien die sich mit der Ökotoxizität von diesen Verbindungen befassen. Gleichwohl lassen diese Studien Rückschlüsse zu, sodass die ökotoxikologische Relevanz von Mykotoxinen in der aquatischen Umwelt, aufgrund der starken Verdünnung in Oberflächengewässern und der niedrigen Exposition, eher als klein einzustufen ist.

CHAPTER 1:

GENERAL INTRODUCTION

1.1 Fate of organic micropollutants

Contamination of natural waters with so called emerging contaminants, including various classes of organic micropollutants, like pharmaceuticals, pesticides, steroids, and personal care products, has gained public interest within the recent years.[1-8] In this context, micropollutants are defined as organic trace pollutants occurring in the nanogram to microgram per liter range, and can be found ubiquitously in the aquatic environment.[9]

Pharmaceuticals, for instance as one class of organic micropollutants, are administered and just partially metabolized in the human organism.[10] Afterwards, these compounds and their metabolites are excreted and enter the aquatic environment via the municipal sewer system, where they are mostly partially removed.[11,12] Veterinary pharmaceuticals, included in various types of manure, and pesticides enter the aquatic environment mainly via run-off and drainage from agricultural fields after their application on the fields. In addition to their beneficial (health) effects, some of those emerging contaminants are rather persistent, and they hold a high bioaccumulation potential. Hence, their input into the aquatic environment via the different mentioned pathways needs to be minimized.

Besides man-made organic micropollutants, also natural occurring contaminants, namely natural toxins, have been recently identified as another class of potential micropollutants.[13-17] Many of those natural toxins are associated with agricultural systems, and due to the intensification and increased monocultures in agricultural production, higher rates of those compounds per area are produced. Additionally, the same emission routes can be presumed for natural toxins, and man-made organic micropollutants, but natural toxins have always contributed to the complex mixture of micropollutants in the aquatic environment.[18]

1.2 Natural toxins

Natural toxins are compounds that are naturally produced by living organisms, e.g., fungi, bacteria, algae, plants and animals.[19] These toxins are not harmful to the organisms themselves, but they may be toxic to others, including animals and humans, when eaten, inhaled, and some already upon dermal contact. Many natural toxins, such as plant hormones, have regulatory roles, while others function as chemical defense factors against attack by pests and pathogens.[20] On the basis of their origin, these toxic metabolites are commonly divided into five main categories: mycotoxins, bacterial toxins, phycotoxins, phytotoxins, and zootoxins.[21]

Mycotoxins, one class of natural toxins, are a diverse and ubiquitous group of fungal metabolites specifically associated with the occurrence of deleterious effects in humans and animals.[22-27] On a global scale, food safety is regularly compromised by the presence of mycotoxins. Hence, their analysis in these matrices has been pushed forward and has regularly been reviewed.[21,28-30]

1.3 Mycotoxin producing fungi

Filamentous fungi infest agricultural commodities, in the field and during storage, as well as organic soil components.[31] A number of fungal genera are able to synthesize mycotoxins, whereof *Alternaria*, *Aspergillus*, *Claviceps*, *Fusarium* and *Penicillium* species (spp.) are the ones of major concern in human and animal health, and produce a plethora of mycotoxins with a great structural diversity, and thus different chemical and physical properties.[32,33] The exact benefit of mycotoxins for fungi is not clearly understood, but their production is associated with the ecology and survival of the organism under different conditions (see also chapter 1.4).[34] The growth and toxin production of fungi may take place at different places; either in the field before harvest or during storage and transportation, under favorable conditions. The intensified, industrialized agriculture promotes infections with mycotoxin producing fungi. Most of these fungi possess characteristics which abet their ubiquitous presence, like a fast growth rate, a high pH tolerance, and a high abundance in the respective environmental compartment.[35]

Mycotoxin producing fungi considered in this doctoral thesis are: *Alternaria, Aspergillus, Claviceps, Fusarium* and *Penicillium*. In the following they will be briefly introduced. Besides *Fusarium* spp., all others can be classified as *Fungi Imperfecti* (Deuteromycotina), because they lack a sexual stage and reproduce by asexual conidia.[36]

The genus *Alternaria* includes both plant-pathogenic and saprophytic species, which may affect crops in the field or cause harvest and postharvest decay of plant products.[37,38] Plants infected with *Alternaria* spp. show spots on fruits, leaves or stems, tuber and fruit rots, or black ears on cereals. Other signs of disease may be blights, collar rot and seedlings damping off. In general, mold growth is accompanied by the production of toxic metabolites like the phytotoxin tentoxin (TEN), and the mycotoxins alternariol (AOH), alternariol monomethyl ether (AME), altenuene (ALT) and tenuazonic acid (TEA).

Aspergillus species infect fruits, vegetables, nuts, bread and hay either in the field or during storage. Within the different *Aspergillus* groups, the *A. niger* group (the black aspergilla) is one of the most widespread food and feed contaminants, and is additionally used by the biotechnical industry for citric acid production.[39] Thus far, despite the long history and intensive nature of *Aspergillus* research, only few cases of toxin formation by *A. niger* have been reported.[40] Asides many other *Aspergillus* strains, *A. fumigatus* and *A. flavus* produce several classes of very toxic mycotoxins, i.e., aflatoxins (AFs), ochratoxins, and citrinin (CIT).[41]

Members of the fungal genus *Claviceps* parasitize economically important crop plants like rye, wheat, barley, rice, corn, millet, and oat.[42] *C. purpurea*, the most widely known species in Europe, is exceptional as it infects more than 400 plant species with a disease known as ergot.[43] The fungus replaces the developing grain or seed with the alkaloid-containing wintering body, called sclerotium.[28] In the medieval times, consumption of ergot-contaminated rye bread caused vast epidemics of the so-called St. Anthony's fire disease. Conversely, ergot sclerotia can induce contractions and were also used by midwives as help in child birth, and till today, this fungal genus has been of considerable commercial interest in the biotechnological production due to their pharmacological effects.[44]

Penicillium spp. is a fungal genus of major importance due to antibiotic properties of penicillin, but also as a ubiquitous food contaminant. Especially, *P. expansum*, the blue mold that induces soft rot of apples, pears, cherries, and other fruits, is recognized as one of the most common producers of mycotoxins. In pure culture, *P. expansum* is reported to produce more than 50 different secondary metabolites, such as citrinin, ochratoxin A (OTA), patulin (PAT), penicillic acid, and roquefortine C.[45-48] PAT and OTA are the most significant of the *Penicillium* metabolites and EU maximum limits were established for certain food products.[49]

Fusarium head blight (FHB) is one of the most devastating fungal cereal diseases in the world, resulting in severe yield, and financial losses,[50,51] and the contamination of grains by several, very potent mycotoxins, that threaten human and animal health.[52-57] Several *Fusarium* species can cause FHB, with *F. graminearum*, *F. poae*, *F. crookwellense*, and *F. avenaceum* being among the most common ones.[58] The distribution of those pathogens causing FHB is reported to be determined partially by climatic factors, mostly temperature and moisture, and agronomic factors, such as soil tillage and special crop rotations.[59-61] The mycotoxins produced by *Fusarium* spp. include the trichothecenes (type A and B), zearalenone (ZON), moniliformin, fumonisins, beauvericin (BEA) and the enniatins, while the trichothecenes form the largest group of *Fusarium* spp. derived mycotoxins.[62]

1.4 Classification and chemical structures of mycotoxins

Mycotoxins are naturally occurring secondary plant metabolites produced by fungi of different genera. Mycotoxins remain challenging to classify due to their diverse chemical structures, biosynthetic origin, and their production by a wide number of fungal genera and species. One approach can be the classification according to differences in the fungal origin, chemical structure and biological activity. The classification can also be done according to their frequency of occurrence and the corresponding detected amounts in each commodity. The last approach is rather complicated, due to the fact, that mycotoxin contamination of food and feed depends on environmental and climatic conditions, harvesting techniques, and storage conditions. Since these secondary plant metabolites have different toxic effects, another classification method is based on the organ, or the system the mycotoxin affects: hepatotoxins, nephrotoxins, neurotoxins, immunotoxins.[63]

In this doctoral thesis, a selection of 33 different and highly complex mycotoxins, originating from five important fungal genera, was investigated. In **Table 1.1** these five fungal genera are listed with several representative species, and a mycotoxin produced together with its chemical structure (for all 33 mycotoxin structures see chapter 2, Figure 2.1). Hence, here a classification is done according to the different genus and the chemical structure.

Table 1.1. Classification of mycotoxin producing fungi.

Genus	Fungal species	Mycotoxins	Chemical structure
	A. flavus, *A. parasiticus*, *A. nomius*	Aflatoxins: **Aflatoxin B_1**, Aflatoxin B_2, Aflatoxin G_1, Aflatoxin G_2, Aflatoxin M_1,	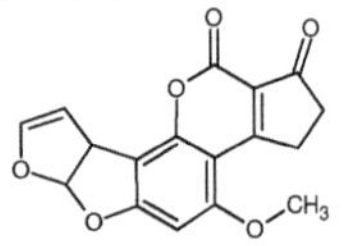
Aspergillus	*A. ochraceus*	Ochratoxins: **Ochratoxin A**, Ochratoxin B	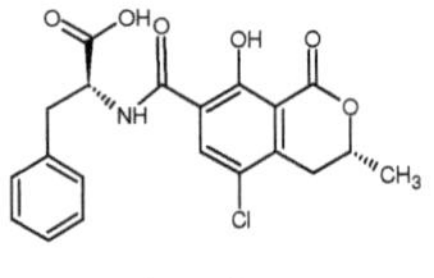
	A. clavatus, *A. terreus*	**Patulin**, Sterigmatocystin, Sulochrin	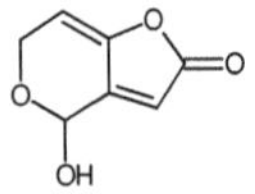

Alternaria	*A. alternata,* *A. solani,* *A. tenuissima,* *A. citri,* *A. longipes*	Alternaria toxins: **Alternariol**, Alternariol monomethylether, Altenuene, Tentoxin	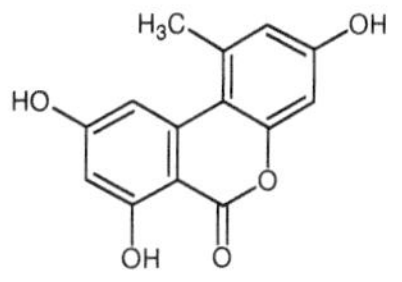
Claviceps	*C. purpurea,* *C. fusiformis,* *C. paspali,* *C. africana*	Ergot alkaloids: **Ergocornine**, Ergocryptine	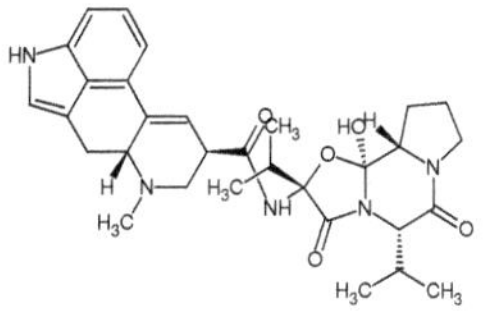
	F. verticillioides, *F. proliferatum,*	Fumonisins: **Fumonisin B_1**, Fumonisin B_2, Beauvericin	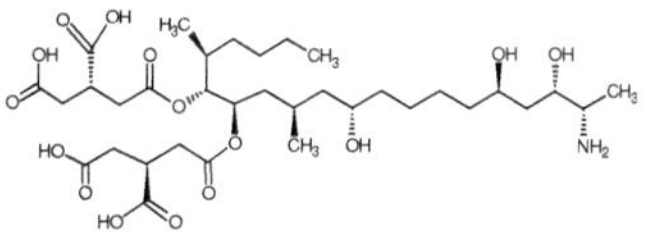
Fusarium	*F. graminearum,* *F. culmorum,* *F. crookwellense,* *F. sporotrichioides,* *F. poae,* *F. acuminatum,* *F. sambucinum*	Type A trichothecenes: Diacetoxyscirpenol ,**HT-2 toxin**, Neosolaniol, T-2 toxin, Verrucarin A, Verrucarol Type B trichothecenes: 3-Acetyl-deoxynivalenol, 15-Acetyl-deoxynivalenol **Deoxynivalenol**, Fusarenone-X, Nivalenol	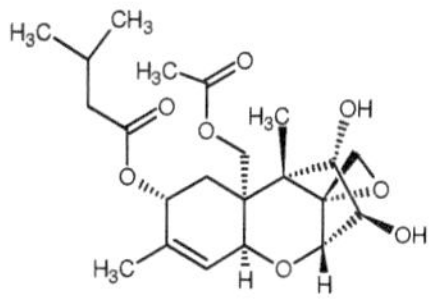 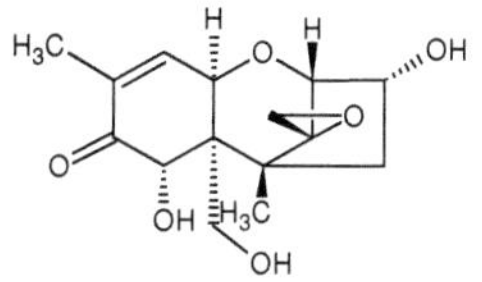
	F. graminearum, *F. culmorum,* *F. avenaceum*	Resorcyclic acid lactones: **Zearalenone**, α-Zearalenone, β-Zearalenone	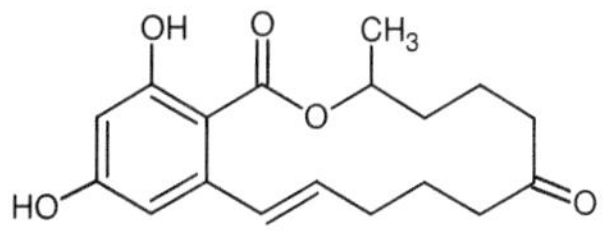
Penicillium	*P. verrucosum,* *P. virridicatum*	Ochratoxins: Ochratoxin A, **Ochratoxin B**	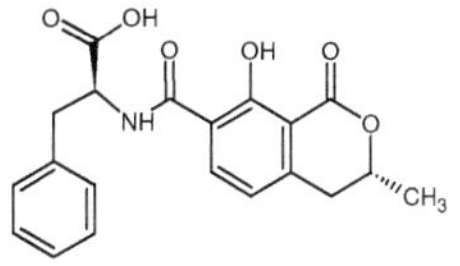
	P. citrinum, *P. verrucosum*	**Citrinin**	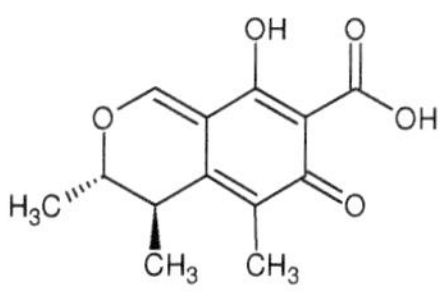

Bold written compound names are depicted with their corresponding chemical structure.

1.5 Occurrence of mycotoxins in food and feed

Globally, mycotoxins can occur in a wide range of commodities, with the highest prevalence being encountered in agricultural products. Here especially small-grain cereals and maize, and respective products, are frequently contaminated, most commonly with aflatoxins (AFs), ochratoxin A (OTA), fumonisins, ZON and trichothecenes (**Table 1.2**). Main factors influencing the prevalence, and the degree of mycotoxin contamination are the geographic region, the climatic conditions, the specific commodity, the harvest year, and storage conditions.[64-67] In Europe, small-grain cereals often contain trichothecenes and ZON.[68,69] In particular, T-2 and HT-2 toxins are becoming more prevalent in the Northern regions of Europe, and especially oats can contain high levels.[70] Additionally, a surveillance of grains and animal feed in Poland documented unacceptable high values for deoxynivalenol (DON), and nivalenol (NIV).[68] Moreover, OTA was repeatedly detected in European crops, although the concentrations were relatively low.[71] Other mycotoxins prevalent especially in Northern European climate include beauvericin (BEA), enniatins (ENNs) and moniliformin (MON).[72,73] Recently published data from Kokkonen et al.[74] demonstrated that also ergot alkaloids can be encountered in grains in Europe, with highest levels reaching the mg/kg level. Additionally, AFs and fumonisins are detected frequently in corn and corn products from the European market. The levels reported for AFs are moderate, but fumonisins have been determined at high concentrations, up to 10.2 mg/kg.[64,75,76]

Other plant products like nuts and oilseeds may contain various types of mycotoxins, including AFs, ZON and *Alternaria* toxins.[38,77-79] Additionally, spices and dried fruits are frequently contaminated with AFs and OTA.[40,80,81] Furthermore, OTA is often found in coffee, tea and cocoa[82,83] and beer.[84,85]

Table 1.2. Some common mycotoxins, commodities in which they are encountered, and concentrations detected in cereal products.

Mycotoxins	Examples of commodities	Detected cereal concentration [µg/kg]	Reference for occurrence data
Aflatoxins:			
AFB_1, AFB_2, AFG_1, AFG_2,	Corn, wheat, rice, spices, sorghum, ground nuts, tree nuts, almonds, oilseeds, dried fruits,	656 (max.), 12-19 (median, AFB_1 in corn and feed)	[86], Reviewed in [64,87,88]
AFM_1	Milk and milk products	Milk mostly < 0.05	[89]
Fumonisins:			
FB_1, FB_2, FB_3	Corn, corn based products, sorghum, asparagus, rice, milk, dried fruits	5-10200 (min/max.), < 0.3-4815 (median)	Reviewed in [30]
Type A trichothecenes:			
HT-2, T-2, DAS, NEO	Cereals, and cereal based products	4-7584 (min/max.), 10-151 (median) for HT-2 and T-2 toxin	Reviewed in [30,90]
Type B trichothecenes:			
3-AcDON, 15- AcDON, NIV, DON, FUS-X	Cereals, and cereal based products	1.7-50000 (min/max.), 1-2000 (median) for DON	Reviewed in [30,90]
ZON	Barley, oats, wheat, rice, sorghum, sesame, soy beans, cereal based products	1-6492 (min/max.), 0.08-689 (median)	Reviewed in [91]
Ochratoxins:			
OTA, OTB	Cereals, dried fruits, wine, coffee, oats, spices, rye, grape juice	0.005-33.33 (min/max.), 0.01-2.0 (median) for OTA	Reviewed in [23],[92]
Ergot alkaloids:			
ECO, EGC, ECY, EGS, EGT	Wheat, rye, barley, millet, oats, sorghum, triticale, cereal based products	2-3280 (min/max.), 1-850 (median) for total ergot alkaloids	[28,93]

AFB_1: aflatoxin B_1, AFB_2: aflatoxin B_2, AFG_1: aflatoxin G_1, AFG_2: aflatoxin G_2, AFM_1: aflatoxin M_1, FB_1: fumonisin B_1, FB_2: fumonisin B_2, FB_3: fumonisin B_3, HT-2: HT-2 toxin, T-2: T-2 toxin, DAS: diacteoxyscirpenol, NEO: neosolaniol, 3-AcDON: 3-acetyl-deoxynivalenol, 15- AcDON: 15-acetly-deoxynivalenol, NIV: nivalenol, DON: deoxynivalenol, FUS-X: fusarenone-x, ZON: zearalenone, OTA: ochratoxin A, OTB: ochratoxin B, ECO: ergocornine, EGC: ergocristine, ECY: ergocryptine, EGS: ergosine; EGT: ergotamine.

Along with the mycotoxins, their metabolites or conjugated forms called masked or bound mycotoxins may also occur in food and feed.[94-96] These compounds can be more soluble or can be associated with macromolecules, and may emerge through the metabolism of fungi or exposed animals as defense mechanism.[97] Some known mycotoxin conjugates are the plant-derived glucosides of ZON and DON. For example, zearalenol 4-glucosides and DON 3-glucoside have been detected in naturally occurring maize and small grain cereals.[94,97,98] Additionally, nivalenol and fusarenone-X form glucosides as well, as recently documented by Nakagawa et al.[99]

In order to protect consumers, regulatory authorities in several countries have set limits for maximum residue levels (MRL) for selected mycotoxins.[21] In the European Union (EU), maximum limits have been established, for instance for the aflatoxins, where in dried fruits or products thereof, the sum of all five aflatoxins-representatives should not exceed 4.0 µg/kg. For example in unprocessed grain, the maximum tolerable amount of OTA is set to 5.0 µg/kg. The maximum tolerable amount of DON in pasta products is set to 750 µg/kg, and for patulin (PAT) in fruit juices to 50 µg/kg. Additionally, the EU established a maximum limit for AFB_1 in feed and feed raw materials.[100-102]

1.6 Toxicity of mycotoxins

Mycotoxins are toxic, either acutely or on chronic exposure. The magnitude of their toxicity depends on several factors, e.g. the level, duration and route of exposure, as well as mode of action, metabolism in the body, and sensitivity of the target animal.[33] Several mycotoxins can impair the immune system of humans and animals, which may consequently lead to exposure to other diseases, and thus mask underlying toxicoses.[103] Mycotoxicoses are generally difficult to diagnose since the biological effects of mycotoxins are diverse and often unspecific, causing symptoms such as nausea, loss of appetite, reduced feed intake and reduced bodyweight.[104]

The toxicological effects of several mycotoxins, such as aflatoxins (AFs), fumonisins, DON, and OTA are well-investigated,[33,105-108] whereas the toxic mechanisms have not been elucidated completely except for certain fungal metabolites, like MON, BEA and ENNs.[109] The toxicity of BEA has been attributed primarily to the antimicrobial and insecticidal property it comprises,[110,111] and the effects demonstrated mostly only at cell level.[112,113]

In general, AFs, fumonisins and OTA are regarded as the most toxic of the known mycotoxins. They are mutagenic and teratogenic, i.e. they may cause alterations in genetic material and impair the development of embryos or fetuses.[114-116] AFs have been demonstrated to cause liver cancer and are classified as human carcinogens belonging to Group 1 according to the International Agency for Research on Cancer.[117,118] Fumonisins and OTA are probable human carcinogens of group 2B.[117,118] Fumonisins have been linked to liver cancer in rats and epidemics of human esophageal cancer, whereas OTA may cause tumors in kidneys and urinary tract.[114,119] Another class of mycotoxins, the resorcyclic acid lactones (RALs), which include ZON, exhibit estrogenic potency,[120] and reproductive disorders of farm animals, and occasionally hyperestrogenic syndromes in humans have been reported.[121]

In vitro tests have also been used for investigating the toxicity of mycotoxin mixtures.[122-126] For example, cytotoxicity has been demonstrated as being additive for OTA, OTB, CIT and PAT in a porcine renal cell line.[124] Recently, a synergistic effect of OTA and FB_1 on the decrease of viability of human and pig lymphocytes was observed.[127] Furthermore, an additive immunotoxic effect of DON and NIV was reported in mouse macrophages.[128] The toxicity of fungal extracts and cereal sample extracts with a known content of mycotoxins has been examined in cell assays.[129-131] These types of studies provide an estimate of the total toxicity of the extract, but it may be difficult to demonstrate the correlation between the toxicity and the mycotoxin concentrations. Furthermore, it is impossible to extrapolate the real toxic effect from these results on animals and humans.

1.7 Occurrence of mycotoxins in the aquatic environment

Madden and co-workers were the first to hypothesize that mycotoxin-contaminated crops are potential contaminants of groundwater.[132] Accordingly, soil column leaching studies for AFs and fumonisins were conducted, showing quite some mobility, especially for the highly water-soluble fumonisins. A similar attempt was followed by Williams and co-workers ten years later solely for fumonisin B_1.[133] Hence, chemical-physical data points are essential values to elucidate the environmental fate and behavior of mycotoxins. In this respect, the octanol/water partition constant (K_{iOW}) and the natural organic-matter (NOM)/water partition constant (K_{iOC}) of a compound, are indicators of the likelihood of a given compound to enter natural water systems. Overall, experimentally determined physical-chemical data points for mycotoxins are rare. Some mechanistic investigations on the sorption behavior of selected mycotoxins have been undertaken by several authors.[134-136] Basically, sorption mechanisms using specific clay minerals were investigated to facilitate possible detoxification of the mycotoxins-contaminated feeds.[136,137] Additionally, humic material has also been considered as mycotoxin binding additive, but so far only for aflatoxin $B_{1,}$[138] DON,[139] OTA,[140] and ZON.[139,140] Besides this, only one experimentally determined NOM-water sorption coefficient for ZON is available so far,[141] but allows for a first estimation of the potential ZON emission from agricultural fields.

Despite these studies, the environmental occurrence and exposure of mycotoxins has scarcely been investigated. Initial studies from our lab on two prominent mycotoxins, i.e. DON and ZON, proved that the aquatic environment can also be exposed to mycotoxins.[13,142] Thus far, the identified main input sources of mycotoxins into the aquatic environment comprise 1) run-off and drainage water from fields cultivated with fungi-infected cereals, 2) husbandry animals (i.e., excretion from grazing livestock, run-off from feeding operations, or manure application), and 3) human excretion via sewer systems.

As indicated above, mycotoxin emission from agricultural fields cultivated with wheat or corn, by run-off and drainage water has recently been investigated in more detail. For ZON, occasional peak concentrations of up to 35 ng/L were reported by Hartmann et al.[142,143] During a follow-up study, DON was detected in drainage water samples with concentrations ranging between 23 ng/L and 4.9 μg/L.[13] These concentrations were considerably higher than those of ZON, which was attributed to the higher amounts of DON produced in the contaminated crop and its higher aqueous solubility. Additionally, it should be emphasized that the mycotoxin amounts produced in the agricultural fields are similar to those of agricultural application rates of many modern pesticides (50 g – 1 kg/ha per whole season).

Thus far, the exposure of husbandry animals to mycotoxins has been well investigated.[144-146] However, mycotoxins are excreted by husbandry animals rather rapidly after the consumption of contaminated feed.[147] Of the administered DON included in the feed, up to 37% was excreted via urine in rats,[148] whereas 3 – 68% of DON and NIV were detected in pig faeces.[144,149] Additionally, Seeling et al.[145] recovered varying fractions of *α*- and *β*-zearalenol (ZOL), and ZON in cow faeces. Moreover, mycotoxins can also enter the aquatic environment via run-off from animal feeding operations. For example, Bartelt-Hunt et al.[150] reported concentrations of *α*-ZOL up to 5.2 μg/L in run-off from cattle feedlots. In this specific case, the ZON metabolites were detected as a result of the application of the growth promoter *α*-zearalanol. Although mycotoxin concentrations in manure have been little investigated so far, two recent studies demonstrated that mycotoxins can be present in manure.[151,152] Hartmann et al.[151] investigated various types of manure and detected ZON in all samples with average concentrations ranging between 50 and 150 ng/g_{dw}. Additionally, Schollenberger et al.[152] found DON in one of seven digested manure samples at the detection limit (20 μg/L), and deepoxy-DON was detected between 6 and 20 μg/L in three of them. Hence, a mycotoxin input in rivers from husbandry animal waste seems likely.

Since humans can be exposed to mycotoxins via food intake, and due to the fact that a certain fraction is usually excreted (native, metabolized, or in conjugated form), the sewer system is a third potentially source of mycotoxins in the aquatic environment. The presence of mycotoxins in WWTPs in- and effluents was already documented by several authors (reviewed in[153]). For instance, the occasional presence of ZON, as well as its metabolites α-ZOL, α-zearalanol, and β-zearalanol in German and Italian WWTP in- and effluents at the low ng/L concentration level was reported by various authors.[154-159] Besides this, Wettstein et al.[160] reported the permanent presence of DON in the primary effluent of three different WWTPs in Switzerland, and concentrations in the WWTP effluent ranged from 18 - 42 ng/L. Calculated elimination rates ranged between 30 - 57% for DON, and attest just a partial elimination during the waste water treatment.

Hence, the above-mentioned sources are probable causes for the exposure of surface waters to mycotoxins. Their occurrence, mostly in the low ng/L-concentration range, has been reported in various surface waters in Europe and the US.[13,154,161,162] However, with concentrations up to 44 ng/L of ZON, the study of Gromadzka et al.[161] is rather exceptional.

1.8 Objectives of this doctoral thesis

Based on the knowledge for DON and ZON presented above, and the fact that comparable or even higher amounts of other mycotoxins are produced in wheat and maize, which exhibit similar or even higher aqueous solubilities, a larger number of mycotoxins is likely to enter the aquatic environment. Hence, the following hypotheses were formulated.

I. Mycotoxins are emitted into the (aqueous) environment from agricultural fields or via waste water treatment plants (WWTP).

II. The relative importance of these input pathways depends on the mycotoxin type, and the produced mycotoxin amount, and whether they are produced in the field or during storage.

III. The magnitude of emission depends on produced amount, persistence, and chemical-physical properties of the individual mycotoxin.

IV. Certain mycotoxins are of relevance in terms of their environmental risk to aquatic ecosystems.

To extend the limited knowledge of environmental fate and behavior of a few compounds to a larger variety of mycotoxins, the following procedures were applied to either accept or reject the above mentioned hypotheses:

(1) Development of an analytical multi-residue screening method for aqueous samples.

(2) Monitoring of mycotoxins in different water matrices (drainage- and surface waters, WWTP-effluent).

(3) For frequently detected mycotoxins in food products and aqueous samples an analytical method for plant samples (whole wheat plants) was developed.

(4) Determination of soil organic-carbon water partitioning coefficients for various mycotoxins.

(5) Establishment of a mycotoxin mass balance on the test field at the ART.

The overall aims of this thesis are to elucidate which mycotoxins occur in natural waters, to pinpoint their major emission pathways, and to assess their ecotoxicological relevance.

1.9 Strategies to achieve the objectives

To achieve these objectives, first of all, analytical methods for the quantification of these compounds in various agro-environmental matrices (aqueous and solid matrices) were needed. Therefore, analytical methods were developed and validated for drainage- and surface waters, as well as for WWTP effluents, and for whole wheat plant samples (chapters 2 and 3).

The environmental fate and behavior is strongly influenced by the physical-chemical properties of the mycotoxins, therefore, a consistent set of experimentally determined sorption coefficients to soil organic matter was generated for selected mycotoxins (chapter 4).

The occurrence of selected mycotoxins on an experimental wheat field, as well as their emission from the field into the drainage water was investigated over nearly two years (chapter 5). Furthermore the analytical method for aqueous matrices was applied to investigate the occurrence of mycotoxins in Swiss rivers and WWTP effluents (chapter 6). The mycotoxin loads for the respective compartments (wheat plants, drainage and surface waters, as well as WWTP effluents) were calculated and discussed in chapter 5 and 6. The emission of mycotoxins from different possible sources (WWTP effluents, and areas with *Fusarium*-infected crops) in surface waters was determined and presented in chapters 5 and 6. Additionally, the ecotoxicological relevance of the presence of these compounds in surface waters was discussed using literature data only (chapter 5-7), but will be assessed and evaluated by data from own ecotoxicological tests in a separate publication.

Further, the occurrence of mycotoxins was also investigated in larger river systems within different agricultural systems (USA, Iowa), were the seasonal pattern was shifted.[162] This fact could be explained by the permanent snow coverage of fields in winter, which were previously cropped with small grain cereals (corn). Results from this survey will also be published separately.

1.10 List of publications out of this doctoral thesis

The following articles are already published, or were submitted to international, peer-reviewed journals:

Paper 1:

Schenzel, J.; Schwarzenbach, R.P.; Bucheli, T.D. Multi-residue screening method to quantify mycotoxins in aqueous environmental samples. *J. Agric. Food Chem.* **2010**, *58*, 11207-11217.

Paper 2:

Schenzel, J.; Forrer, H.-R.; Vogelgsang, S.; Bucheli, T.D. Development, validation and application of a multi-mycotoxin method for the analysis of whole wheat plants. *Mycotoxin Res.* **2012**, *28*, 135-147.

Paper 3:

Schenzel, J.; Bucheli T.D.; Goss K.-U.; Schwarzenbach, R.P.; Droge, S.T.J. Experimentally determined soil organic matter-water partitioning coefficients for different classes of natural toxins and comparison with estimated numbers. *Environ. Sci. Technol.* **2012**, *46*, 6118-6126.

Paper 4:

Schenzel, J.; Forrer, H.-R.; Vogelgsang, S.; Hungerbühler, K.; Bucheli T.D. Mycotoxins in the Environment: I. Production and emission from an agricultural test field. *Environ. Sci. Technol.* **2012**, *46*, 13067-75.

Paper 5:

Schenzel, J.; Hungerbühler, K.; Bucheli T.D. Mycotoxins in the Environment: II. Occurrence and origin of mycotoxins in Swiss river waters. *Environ. Sci. Technol.* **2012**, *46*, 13076-84.

References (chapter 1)

1. Byrns, G., The fate of xenobiotic organic compounds in wastewater treatment plants. *Wat. Res.* **2001,** *35*, 2523-2533.
2. Freitas, L. G., et al., Quantification of the new triketone herbicides, sulcotrione and mesotrione, and other important herbicides and metabolites, at the ng/l level in surface waters using liquid chromatography- tandem mass spectrometry. *J. Chromatogr. A* **2004,** *1028*, 277-286.
3. Kasprzyk-Hordern, B., et al., The effect of signal suppression and mobile phase composition on the simultaneous analysis of multiple classes of acidic/neutral pharmaceuticals and personal care products in surface water by solid-phase extraction and ultra performance liquid chromatography-negative electrospray tandem mass spectrometry. *Talanta* **2008,** *74*, 1299-1312.
4. McArdell, C. S., et al., Occurrence and fate of macrolide antibiotics in wastewater treatment plants and in the Glatt Valley Watershed, Switzerland. *Environ. Sci. Technol.* **2003,** *37*, 5479-5486.
5. Reemtsma, T.; Quintana, J. B., Analytical Method for Polar Pollutants In *Organic Pollutants in the Water Cycle*, Reemtsma, T.; Jekel, M., Eds. Wiley-VCH Verlag GmbH & Co. KGaA: Weinheim, 2006; Vol. 1, pp 1-34.
6. Schwarzenbach, R. P., et al., The challenge of micropollutants in aquatic systems. *Science* **2006,** *313*, 1072-1077.
7. Thomas, P. M.; Foster, G. D., Tracking acidic pharmaceuticals, caffeine, and triclosan through the wastewater treatment process. *Environ. Toxicol. Chem.* **2005,** *24*, 25-30.
8. Vanderford, B. J., et al., Analysis of endocrine disruptors, pharmaceuticals, and personal care products in water using liquid chromatography/tandem mass spectrometry. *Anal. Chem.* **2003,** *75*, 6265-6274.
9. Schwarzenbach, R. P., et al., Global water pollution and human health. In *Ann. Rev. Environ. Resour. Vol 35*, 2010; pp 109-136.
10. Halling-Sorensen, B., et al., Occurrence, fate and effects of pharmaceutical substances in the environment - A review. *Chemosphere* **1998,** *36*, 357-394.

11. Hollender, J., et al., Elimination of organic micropollutants in a municipal wastewater treatment plant upgraded with a full-scale post-ozonation followed by sand filtration. *Environ. Sci. Technol.* **2009,** *43*, 7862-7869.
12. Joss, A., et al., Removal of pharmaceuticals and fragrances in biological wastewater treatment. *Wat. Res.* **2005,** *39*, 3139-3152.
13. Bucheli, T. D., et al., *Fusarium* mycotoxins: Overlooked aquatic micropollutants? *J. Agric. Food Chem.* **2008,** *56*, 1029-1034.
14. Hartmann, N., et al., Occurrence of zearalenone on *Fusarium graminearum* infected wheat and maize fields in crop organs, soil, and drainage water. *Environ. Sci. Technol.* **2008,** *42*, 5455-5460.
15. Jensen, P. H., et al., Extraction and determination of the potato glycoalkaloid alpha-solanine in soil. *Inter. J. Environ. Anal. Chem.* **2007,** *87*, 813-824.
16. Rasmussen, L. H., et al., Distribution of the carcinogenic terpene ptaquiloside in bracken fronds, rhizomes (*Pteridium aquilinum*), and litter in Denmark. *J. Chem. Ecol.* **2003,** *29*, 771-778.
17. Rasmussen, L. H., et al., Occurrence of the carcinogenic Bracken constituent ptaquiloside in fronds, topsoils and organic soil layers in Denmark. *Chemosphere* **2003,** *51*, 117-127.
18. Hoerger, C. C., et al., Analysis of selected phytotoxins and mycotoxins in environmental samples. *Anal. Bioanal. Chem.* **2009,** *395*, 1261-1289.
19. Rimado, A. M.; Duke, S. O., Natural products for pest management. In *Selected Topics in the Chemistry of Natural Products*, Ikan, R., Ed. 2007; pp 209-251.
20. Ikan, R., The Origin and the nature of natural products. In *Selected Topics - In the Chemistry of Natural Products*, Ikan, R., Ed. World Scientific Publishing Co. Pte. Ltd.: Danvers, MA, U.S.A., 2008; pp 1-9.
21. van Egmond, H. P., Natural toxins: risks, regulations and the analytical situation in Europe. *Anal. Bioanal. Chem.* **2004,** *378*, 1152-1160.
22. Torres-Sanchez, L.; Lopez-Carrillo, L., Fumonisin intake and human health. *Salud Publica De Mexico* **2010,** *52*, 461-467.
23. Reddy, K. R. N., et al., An overview of mycotoxin contamination in foods and its implications for human health. *Toxin Rev.* **2010,** *29*, 3-26.
24. Lairon, D., Nutritional quality and safety of organic food. A review. *Agron. Sustain. Dev.* **2010,** *30*, 33-41.

25. Muri, S. D., et al., Comparison of human health risks resulting from exposure to fungicides and mycotoxins via food. *Food Chem. Toxicol.* **2009,** *47*, 2963-2974.
26. Shephard, G. S., Impact of mycotoxins on human health in developing countries. *Food Add. Contam.* **2008,** *25*, 146-151.
27. Fung, F.; Clark, R. F., Health effects of mycotoxins: A toxicological overview. *J. Toxico.-Clinical Toxicol.* **2004,** *42*, 217-234.
28. Krska, R.; Crews, C., Significance, chemistry and determination of ergot alkaloids: A review. *Food Add. Contam.* **2008,** *25*, 722-731.
29. Turner, N. W., et al., Analytical methods for determination of mycotoxins: A review. *Anal. Chim. Acta* **2009,** *632*, 168-180.
30. Zollner, P.; Mayer-Helm, B., Trace mycotoxin analysis in complex biological and food matrices by liquid chromatography-atmospheric pressure ionisation mass spectrometry. *J. Chromatogr. A* **2006,** *1136*, 123-169.
31. Bhat, R., et al., Mycotoxins in Food and Feed: Present Status and Future Concerns. *Compr. Rev. Food. Sci. Food Saf.* **2010,** *9*, 57-81.
32. Bennett, J. W.; Klich, M., Mycotoxins. *Clin. Microbiol. Rev.* **2003,** *16*, 497-516.
33. Hussein, H. S.; Brasel, J. M., Toxicity, metabolism, and impact of mycotoxins on humans and animals. *Toxicology* **2001,** *167*, 101-134.
34. Magan, N.; Aldred, D., Post-harvest control strategies: Minimizing mycotoxins in the food chain. *Int. J. Food Microbiol.* **2007,** *119*, 131-139.
35. Nielsen, K. F., et al., Review of secondary metabolites and mycotoxins from the *Aspergillus niger* group. *Anal. Bioanal. Chem.* **2009,** *395*, 1225-1242.
36. Asam, S., et al., Precise determination of the *Alternaria* mycotoxins alternariol and alternariol monomethyl ether in cereal, fruit and vegetable products using stable isotope dilution assays. *Mycotoxin Res.* **2011,** *27*, 23-28.
37. Thomma, B., *Alternaria* spp.: from general saprophyte to specific parasite. *Mol. Plant Pathol.* **2003,** *4*, 225-236.
38. Logrieco, A., et al., *Alternaria* toxins and plant diseases: an overview of origin, occurrence and risks. *World Mycotoxin J.* **2009,** *2*, 129-140.
39. Schuster, E., et al., On the safety of *Aspergillus niger* - a review. *Appl. Microbiol. Biotechnol.* **2002,** *59*, 426-435.
40. Knudsen, P. B., et al., Occurrence of furnonisins B(2) and B(4) in retail raisins. *J. Agric. Food Chem.* **2011,** *59*, 772-776.

41. Accinelli, C., et al., *Aspergillus flavus* aflatoxin occurrence and expression of aflatoxin biosynthesis genes in soil. *Can. J. Microbiol.* **2008,** *54*, 371-379.
42. Lorenz, N., et al., The ergot alkaloid gene cluster: Functional analyses and evolutionary aspects. *Phytochem.* **2009,** *70*, 1822-1832.
43. Schardl, C. L., et al., Ergot alkaloids - biology and molecular biology. *Alkaloids* **2006**.
44. Schardl, C. L., et al., Chapter 2 Ergot Alkaloids - Biology and Molecular Biology. In *The Alkaloids: Chemistry and Biology*, Academic Press: 2006; Vol. Volume 63, pp 45-86.
45. Brian, P. W., et al., Production of patulin in apple fruits by *Penicillium expansum*. *Nature* **1956,** *178*, 263-264.
46. Bridge, P. D., et al., A Reappraisal of the terverticillate *Penicillia* using biochemical, physiological and morphological features .1. Numerical Taxonomy. *J. Gen. Microbiol.* **1989,** *135*, 2941-2966.
47. Paterson, R. R. M., et al., Mycopesticidal effects of characterized extracts of *Penicillium* isolates and purified secondary metabolites (including mycotoxins) on *Drosophila melanogaster* and *Spodoptora littoralis*. *J. Invertebr. Pathol.* **1987,** *50*, 124-133.
48. Frisvad, J. C.; Filtenborg, O., Classification of terverticillate *Penicillia* based on profiles of myco-toxins and other secondary metabolites. *Appl. Environ. Microbiol.* **1983,** *46*, 1301-1310.
49. Paterson, R. R. M., et al., Solutions to Penicillium taxonomy crucial to mycotoxin research and health. *Res. Microbiol.* **2004,** *155*, 507-513.
50. Windels, C. E., Economic and Social Impacts of Fusarium Head Blight: Changing Farms and Rural Communities in the Northern Great Plains. *Phytopathol.* **2000,** *90*, 17-21.
51. Nganje, W. E., et al., Regional economic impacts of Fusarium Head Blight in wheat and barley. *Rev. Agric. Econ.* **2004,** *26*, 332-347.
52. Buck, H. T., et al., *Wheat production in stressed environments*. Springer: Dordrecht, The Netherlands, 2007; Vol. 1.
53. Cole, R. J., et al., *Handbook of secondary fungal metabolites*. Academic Press: New York: 2003; Vol. III.

54. Desjardin, A. E., *Fusarium Mycotoxins - Chemistry, Genetics and Biology*. 1. ed.; The American Phytopathological Society: St. Paul, Minnesota 55121, U.S.A., 2006; Vol. 1, p 240.

55. Mücke, W.; Lemmen, C., *Schimmelpilze - Vorkommen, Gesundheitsgefahren, Schutzmassnahmen*. ecomed Medizin, Verlagsgruppe Hüthig Jehle Rehm GmbH: 2004; Vol. 3., überarbeitete und erweiterte Auflage, p 182.

56. Richard, J. L., et al., *Mycotoxins: Risk in Plant, Animal, and Human Systems*. Library of Congress Cataloging in Publication Data: Ames, Iowa, U.S.A., 2003; p 191.

57. Ueno, Y., *Trichothecenes - Chemical, Biological and Toxicological Aspects*. Kodansha/Elsevier: Tokio/Amsterdam, 1983; p 313.

58. Nielsen, L. K., et al., Fusarium Head Blight of Cereals in Denmark: Species Complex and Related Mycotoxins. *Phytopathol.* **2011,** *101*, 960-969.

59. Champeil, A., et al., Fusarium head blight: epidemiological origin of the effects of cultural practices on head blight attacks and the production of mycotoxins by *Fusarium* in wheat grains. *Plant Science* **2004,** *166*, 1389-1415.

60. Xu, X. M.; Nicholson, P., Community Ecology of Fungal Pathogens Causing Wheat Head Blight. In *Ann. Rev. Phytopathol.*, 2009; Vol. 47, pp 83-103.

61. Dill-Macky, R.; Jones, R. K., The effect of previous crop residues and tillage on Fusarium head blight of wheat. *Plant Disease* **2000,** *84*, 71-76.

62. Stepien, L.; Chelkowski, J., Fusarium head blight of wheat: pathogenic species and their mycotoxins. *World Mycotoxin J.* **2010,** *3*, 107-119.

63. Coppock, R. W.; Jacobsen, B. J., Mycotoxins in animal and human patients. *Toxicol. Ind. Health.* **2009,** *25*, 637-655.

64. Binder, E. M., et al., Worldwide occurrence of mycotoxins in commodities, feeds and feed ingredients. *Ani. Feed Sci.Technol.* **2007,** *137*, 265-282.

65. Scudamore, K. A.; Patel, S., Occurrence of *Fusarium* mycotoxins in maize imported into the UK, 2004-2007. *Food Addit. Contam. Part A-Chem.* **2009,** *26*, 363-371.

66. Scudamore, K. A., et al., HT-2 toxin and T-2 toxin in commercial cereal processing in the United Kingdom, 2004-2007. *World Mycotoxin J.* **2009,** *2*, 357-365.

67. Scudamore, K. A., et al., *Fusarium* mycotoxins in the food chain: maize-based snack foods. *World Mycotoxin J.* **2009,** *2*, 441-450.

68. Placinta, C. M., et al., A review of worldwide contamination of cereal grains and animal feed with *Fusarium* mycotoxins. *Ani. Feed Sci.Technol.* **1999,** *78*, 21-37.
69. Schollenberger, M., et al., Survey of Fusarium toxins in foodstuffs of plant origin marketed in Germany. *Int. J. Food Microbiol.* **2005,** *97*, 317-326.
70. Edwards, S. G., et al., Emerging issues of HT-2 and T-2 toxins in European cereal production. *World Mycotoxin J.* **2009,** *2*, 173-179.
71. Miraglia, M., et al., Climate change and food safety: An emerging issue with special focus on Europe. *Food Chem. Toxicol.* **2009,** *47*, 1009-1021.
72. Uhlig, S.; Ivanova, L., Determination of beauvericin and four other enniatins in grain by liquid chromatography-mass spectrometry. *J. Chromatogr. A* **2004,** *1050*, 173-178.
73. Uhlig, S., et al., Moniliformin in Norwegian grain. *Food Add. Contam.* **2004,** *21*, 598-606.
74. Kokkonen, M.; Jestoi, M., Determination of ergot alkaloids from grains with UPLC-MS/MS. *J. Sep. Sci.* **2010,** *33*, 2322-2327.
75. Sydenham, E. W., et al., Liquid chromatographic determination of fumonisins B-1, B-2, and B-3 in Corn: Aoac-Iupac Collaborative Study. *Journal of Aoac International* **1996,** *79*, 688-696.
76. D'Arco, G., et al., Analysis of fumonisins B-1, B-2 and B-3 in corn-based baby food by pressurized liquid extraction and liquid chromatography/tandem mass spectrometry. *J. Chromatogr. A* **2008,** *1209*, 188-194.
77. Schollenberger, M., et al., Natural occurrence of 16 *Fusarium* toxins in edible oil marketed in Germany. *Food Control* **2008,** *19*, 475-482.
78. Schollenberger, M., et al., Natural occurrence of *Fusarium* toxins in soy food marketed in Germany. *Int. J. Food Microbiol.* **2007,** *113*, 142-146.
79. Schollenberger, M., et al., Trichothecene toxins in different groups of conventional and organic bread of the German market. *J. Food Comp. Anal.* **2005,** *18*, 69-78.
80. Spanjer, M. C., et al., LC-MS/MS multi-method for mycotoxins after single extraction, with validation data for peanut, pistachio, wheat, maize, cornflakes, raisins and figs. *Food Add. Contam.* **2008,** *25*, 472-489.
81. Monbaliu, S., et al., Development of a multi-mycotoxin liquid chromatography/tandem mass spectrometry method for sweet pepper analysis. *Rapid Commun. Mass Spectrom.* **2009,** *23*, 3-11.

82. Batista, L. R., et al., Ochratoxin A in coffee beans (Coffea arabica L.) processed by dry and wet methods. *Food Control* **2009,** *20*, 784-790.

83. Bacaloni, A., et al., Automated on-line solid-phase extraction-liquid chromatography-electrospray tandem mass spectrometry method for the determination of ochratoxin A in wine and beer. *J. Agric. Food Chem.* **2005,** *53*, 5518-5525.

84. Romero-Gonzalez, R., et al., Application of conventional solid-phase extraction for multimycotoxin analysis in beers by ultrahigh-performance liquid chromatography-tandem mass spectrometry. *J. Agric. Food Chem.* **2009,** *57*, 9385-9392.

85. Zachariasova, M., et al., Deoxynivalenol and its conjugates in beer: A critical assessment of data obtained by enzyme-linked immunosorbent assay and liquid chromatography coupled to tandem mass spectrometry. *Anal. Chim. Acta* **2008,** *625*, 77-86.

86. Shephard, G. S., Determination of mycotoxins in human foods. *Chem. Soc. Rev.* **2008,** *37*, 2468-2477.

87. Reiter, E., et al., Review on sample preparation strategies and methods used for the analysis of aflatoxins in food and feed. *Mol. Nutr. Food Res.* **2009,** *53*, 508-524.

88. Rustom, I. Y. S., Aflatoxin in food and feed: Occurrence, legislation and inactivation by physical methods. *Food Chem.* **1997,** *59*, 57-67.

89. Wang, H., et al., Determination of aflatoxin M1 in milk by triple quadrupole liquid chromatography-tandem mass spectrometry. *Food Addit. Contam. Part A-Chem.* **2010,** *27*, 1261-1265.

90. Murphy, P. A., et al., Food mycotoxins: An update. *J. Food Sci.* **2006,** *71*, R51-R65.

91. Maragos, C. M., Zearalenone occurrence and human exposure. *World Mycotoxin J.* **2010,** *3*, 369-383.

92. Hohler, D., Ochratoxin A in food and feed: occurrence, legislation and mode of action. *Z. Ernahrung.* **1998,** *37*, 2-12.

93. Koleva, II, et al., Alkaloids in the human food chain - Natural occurrence and possible adverse effects. *Mol. Nutr. Food Res.* **2012,** *56*, 30-52.

94. Berthiller, F., et al., Masked mycotoxins: Determination of a deoxynivalenol glucoside in artificially and naturally contaminated wheat by liquid

chromatography-tandem mass spectrometry. *J. Agric. Food Chem.* **2005,** *53*, 3421-3425.

95. Vendl, O., et al., Occurrence of free and conjugated *Fusarium* mycotoxins in cereal-based food. *Food Addit. Contam. Part A-Chem.* **2010,** *27*, 1148-1152.
96. Vendl, O., et al., Simultaneous determination of deoxynivalenol, zearalenone, and their major masked metabolites in cereal-based food by LC-MS-MS. *Anal. Bioanal. Chem.* **2009,** *395*, 1347-1354.
97. Berthiller, F., et al., Formation, determination and significance of masked and other conjugated mycotoxins. *Anal. Bioanal. Chem.* **2009,** *395*, 1243-1252.
98. Schneweis, I., et al., Occurrence of Zearalenone-4-Beta-D-Glucopyranoside in Wheat. *J. Agric. Food Chem.* **2002,** *50*, 1736-1738.
99. Nakagawa, H., et al., Detection of a new *Fusarium* masked mycotoxin in wheat grain by high-resolution LC-Orbitrap (TM) MS. *Food Addit. Contam. Part A-Chem.* **2011,** *28*, 1447-1456.
100. E.U., Directive 2002/32/EC of the European Parliament and of the Council of 7 May 2002 on undesirable substances in animal feed. In Journal of the European Communities 2002; Vol. 140, pp 10-22.
101. E.U., Commission Regulation (EC) No 1881/2006 of 19 December 2006 setting maximum levels for certain contaminants in foodstuffs. In Official Journal of the European Communities 2006; Vol. 364, pp 5-24.
102. E.U., Commission Regulation (EC) No 1126/2007 of 28 September 2007 amending Regulation (EC) No 1881/2006 setting maximum levels for certain contaminants in foodstuffs as regards Fusarium toxins in maize and maize products. In Official Journal of the European Communities: 2007; Vol. 255, pp 14-17.
103. Bennett, J. W., Mycotoxins, mycotoxicoses, mycotoxicology and mycopathologia. *Mycopath.* **1987,** *100*, 3-5.
104. Schlatter, J. In *Toxicity data relevant for hazard characterization*, Workshop on Trichothecenes with a Special Focus on DON, Dublin, IRELAND, Sep 10-12, 2003; Elsevier Sci Ireland Ltd: Dublin, IRELAND, 2003; pp 83-89.
105. Stockmann-Juvala, H.; Savolainen, K., A review of the toxic effects and mechanisms of action of fumonisin B-1. *Hum. Exper. Toxicol.* **2008,** *27*, 799-809.

106. Iverson, F., et al., Chronic feeding study of deoxynivalenol in B6C3F1 male and female mice. *Terato. Carcin. Mut.* **1995,** *15*, 283-306.

107. Pestka, J. J.; Smolinski, A. T., Deoxynivalenol: Toxicology and potential effects on humans. *J. Toxicol. Environ. Heal.-Part B-Critical Reviews* **2005,** *8*, 39-69.

108. Orsi, R. B., et al., Acute toxicity of a single gavage dose of fumonisin B-1 in rabbits. *Chemico-Biological Interactions* **2009,** *179*, 351-355.

109. Jestoi, M., et al., Analysis of the *Fusarium* mycotoxins fusaproliferin and trichothecenes in grains using gas chromatography-mass spectrometry. *J. Agric. Food Chem.* **2004,** *52*, 1464-1469.

110. Grove, J. F.; Pople, M., The Insecticidal Activity of Beauvericin and the Enniatin Complex. *Mycopathologia* **1980,** *70*, 103-105.

111. Gupta, S., et al., Isolation of novel beauvericin analogs from the fungus *Beauveria bassiana. Journal of Natural Products-Lloydia* **1995,** *58*, 733-738.

112. Fotso, J.; Smith, J. S., Evaluation of beauvericin toxicity with the bacterial bioluminescence assay and the Ames mutagenicity bioassay. *J. Food Sci.* **2003,** *68*, 1938-1941.

113. Jestoi, M., Emerging *Fusarium*-mycotoxins fusaproliferin, beauvericin, enniatins, and moniliformin - A review. *Critical Reviews in Food Sci. Nutri.* **2008,** *48*, 21-49.

114. Pfohl-Leszkowicz, A.; Manderville, R. A., An update on direct genotoxicity as a molecular mechanism of ochratoxin A carcinogenicity. *Chem. Res. Toxicol.* **2012,** *25*, 252-262.

115. Hadjeba-Medjdoub, K., et al., Structure-Activity Relationships imply different mechanisms of action for ochratoxin A-mediated cytotoxicity and genotoxicity. *Chem. Res. Toxicol.* **2012,** *25*, 181-190.

116. Pfohl-Leszkowicz, A., et al., Combined toxic effects of ochratoxin A and citrinin, in vivo and in vitro. In *Food Contaminants: Mycotoxins and Food Allergens*, Siantar, D. P.; Trucksess, M. W.; Scott, P. M.; Herman, E. M., Eds. Amer Chemical Soc: Washington, 2008; Vol. 1001, pp 56-79.

117. IARC. *Some naturally occurring substances: food items and constituents, heterocyclic aromatic amines and mycotoxins, No. 56, Ochratoxin A.*; International Agency for Research on Cancer: 1993; pp 489-521.

118. IARC. *Monographs on the evaluation of carcinogenic risks to humans: Some traditional herbal medicines, some mycotoxins, naphthalene and styrene*; International Agency for Research on Cancer: Lyon, 2002.

119. van der Westhuizen, L., et al., Individual fumonisin exposure and sphingoid base levels in rural populations consuming maize in South Africa. *Food Chem. Toxicol.* **2010,** *48*, 1698-1703.

120. Zinedine, A., et al., Review on the toxicity, occurrence, metabolism, detoxification, regulations and intake of zearalenone: An oestrogenic mycotoxin. *Food Chem. Toxicol.* **2007,** *45*, 1-18.

121. Metzler, M., et al., Zearalenone and its metabolites as endocrine disrupting chemicals. *World Mycotoxin J.* **2010,** *3*, 385-401.

122. Thuvander, A., et al., In vitro exposure of human lymphocytes to trichothecenes: Individual variation in sensitivity and effects of combined exposure on lymphocyte function. *Food Chem. Toxicol.* **1999,** *37*, 639-648.

123. Tajima, O., et al., Statistically designed experiments in a tiered approach to screen mixtures of Fusarium mycotoxins for possible interactions. *Food Chem. Toxicol.* **2002,** *40*, 685-695.

124. Heussner, A. H., et al., In vitro investigation of individual and combined cytotoxic effects of ochratoxin A and other selected mycotoxins on renal cells. *Toxicol. Vitro* **2006,** *20*, 332-341.

125. Tammer, B., et al., Combined effects of mycotoxin mixtures on human T cell function. *Toxicol. Lett.* **2007,** *170*, 124-133.

126. Speijers, G. J. A.; Speijers, M. H. M., Combined toxic effects of mycotoxins. *Toxicol. Lett.* **2004,** *153*, 91-98.

127. Mwanza, M., et al., The cytotoxic effect of fumonisin B 1 and ochratoxin A on human and pig lymphocytes using the Methyl Thiazol Tetrazolium (MTT) assay. *Mycotoxin Res.* **2009,** *25*, 233-238.

128. Sugiyama, K., et al., Effect of a combination of deoxynivalenol and nivalenol on lipopolisaccharide-induced nitric oxide production by mouse macrophages. *Mycotoxin Res.* **2011,** *27*, 57-62.

129. Gutleb, A. C., et al., Cytotoxicity assays for mycotoxins produced by Fusarium strains: a review. *Environ. Toxicol. Pharmacol.* **2002,** *11*, 309-320.

130. Jestoi, M., et al., Levels of mycotoxins and sample cytotoxicity of selected organic and conventional grain-based products purchased from Finnish and Italian markets. *Mol. Nutr. Food Res.* **2004,** *48*, 299-307.

131. Uhlig, S., et al., Multiple regression analysis as a tool for the identification of relations between semi-quantitative LC-MS data and cytotoxicity of extracts of the fungus *Fusarium avenaceum* (syn. *F-arthrosporioides*). *Toxicon* **2006,** *48*, 567-579.

132. Madden, U. A.; Stahr, H. M., Preliminary determination of mycotoxin binding to soil when leaching through soil with water. *Int. Biodeter. Biodegra.* **1993,** *31*, 265-275.

133. Williams, L. D., et al., Leaching and binding of fumonisins in soil microcosms. *J. Agric. Food Chem.* **2003,** *51*, 685-690.

134. Lemke, S. L., et al., Adsorption of zearalenone by organophilic montmorillonite Clay. *J. Agric. Food Chem.* **1998,** *46*, 3789-3796.

135. Dakovic, A., et al., T-2 toxin adsorption by hectorite. *J. Serb. Chem. Soc.* **2009,** *74*, 1283-1292.

136. Deng, Y. J., et al., Bonding mechanisms between aflatoxin B-1 and smectite. *Appl. Clay Sci.* **2010,** *50*, 92-98.

137. Phillips, T. D., et al., Reducing human exposure to aflatoxin through the use of clay: A review. *Food Add. Contam.* **2008,** *25*, 134-145.

138. van Rensburg, C. J., et al., In vitro and in vivo assessment of humic acid as an aflatoxin binder in broiler chickens. *Poultry Science* **2006,** *85*, 1576-1583.

139. Sabater-Vilar, M., et al., In vitro assessment of adsorbents aiming to prevent deoxynivalenol and zearalenone mycotoxicoses. *Mycopathologia* **2007,** *163*, 81-90.

140. Santos, R. R., et al., Isotherm modeling of organic activated bentonite and humic acid polymer used as mycotoxin adsorbents. *Food Addit. Contam. Part A-Chem.* **2011,** *28*, 1578-1589.

141. Hartmann, N. Environmental exposure to estrogenic mycotoxins. Swiss Federal Institute of Technology, Zurich, 2008.

142. Hartmann, N., et al., Occurrence of zearalenone on *Fusarium graminearum* infected wheat and maize fields in crop organs, soil and drainage water. *Environ. Sci. Technol.* **2008,** *42*, 5455-5460.

143. Hartmann, N., et al., Quantification of estrogenic mycotoxins at the ng/L level in aqueous environmental samples using deuterated internal standards. *J. Chromatogr. A* **2007,** *1138*, 132-140.

144. Danicke, S., et al., On the toxicokinetics and the metabolism of deoxynivalenol (DON) in the pig. *Archi. Ani. Nutri.* **2004,** *58*, 169-180.

145. Seeling, K., et al., On the effects of *Fusarium* toxin-contaminated wheat and the feed intake level on the metabolism and carry over of zearalenone in dairy cows. *Food Add. Contam.* **2005,** *22*, 847-855.

146. Danicke, S., et al., Effects of *Fusarium* toxin-contaminated wheat grain on nutrient turnover, microbial protein synthesis and metabolism of deoxynivalenol and zearalenone in the rumen of dairy cows. *J. Ani. Physiol. Ani. Nutri.* **2005,** *89*, 303-315.

147. Cavret, S.; Lecoeur, S., Fusariotoxin transfer in animal. *Food Chem. Toxicol.* **2006,** *44*, 444-453.

148. Meky, F. A., et al., Development of a urinary biomarker of human exposure to deoxynivalenol. *Food Chem. Toxicol.* **2003,** *41*, 265-273.

149. Hedman, R., et al., Absorption and metabolism of nivalenol in pigs. *Archi. Ani. Nutri.* **1997,** *50*, 13-24.

150. Bartelt-Hunt, S. L., et al., Effect of growth promotants on the occurrence of endogenous and synthetic steroid hormones on feedlot soils and in runoff from beef cattle feeding operations. *Environ. Sci. Technol.* **2012,** *46*, 1352-1360.

151. Hartmann, N., et al., Quantification of zearalenone in various solid agroenvironmental samples using D-6-zearalenone as the internal standard. *J. Agric. Food Chem.* **2008,** *56*, 2926-2932.

152. Schollenberger, M., et al., Simultaneous determination of a spectrum of trichothecene toxins out of residuals of biogas production. *J. Chromatogr. A* **2008,** *1193*, 92-96.

153. Bester, K., et al., Quantitative Mass Flows Of Selected Xenobiotics In Urban Waters And Waste Water Treatment Plants. In *Enviornmental Pollution* Fatta-Kassinos, D.; Bester, K.; Kuemmerer, K., Eds. Springer Netherlands: Dordrecht, 2010; Vol. 16.

154. Lagana, A., et al., Analytical methodologies for determining the occurrence of endocrine disrupting chemicals in sewage treatment plants and natural waters. *Anal. Chim. Acta* **2004,** *501*, 79-88.

155. Lagana, A., et al., Development of an analytical system for the simultaneous determination of anabolic macrocyclic lactones in aquatic environmental samples. *Rapid Commun. Mass Spectrom.* **2001,** *15*, 304-310.

156. Pawlowski, S., et al., Estrogenicity of solid phase-extracted water samples from two municipal sewage treatment plant effluents and river Rhine water using the yeast estrogen screen. *Toxicol. Vitro* **2004,** *18*, 129-138.

157. Spengler, P., Identifizierung und Quantifizierung von Verbindungen mit östrogener Wirkung im Abwasser. *Ph. D. Thesis. Institut für Siedlungswasserbau, Wassergüte und Abfallwirtschaft. Universität Stuttgart. Germany.* **2001**.

158. Spengler, P., et al., Substances with estrogenic activity in effluents of sewage treatment plants in southwestern Germany. 1. Chemical analysis. *Environ. Toxicol. Chem.* **2001,** *20*, 2133-2141.

159. Ternes, T. A., et al. *Pflanzliche Hormonell Wirksame Stoffe in der Aquatischen Umwelt und deren Verhalten bei der Trinkwasseraufbereitung*; Forschungsvorhaben: 02-WU9735/1; ESWE-Institut für Wasserforschung und Wassertechnologie GmbH: 2001; p 114.

160. Wettstein, F. E.; Bucheli, T. D., Poor elimination rates in waste water treatment plants lead to continuous emission of deoxynivalenol into the aquatic environment. *Wat. Res.* **2010,** *44*, 4137-4142.

161. Gromadzka, K., et al., Occurrence of estrogenic mycotoxin - Zearalenone in aqueous environmental samples with various NOM content. *Wat. Res.* **2009,** *43*, 1051-1059.

162. Kolpin, D. W., et al., Phytoestrogens and mycotoxins in Iowa streams: An examination of underinvestigated compounds in agricultural basins. *J. Environ. Qual.* **2010,** *39*, 2089-2099.

CHAPTER 2:

MULTI-RESIDUE SCREENING METHOD TO QUANTIFY MYCOTOXINS IN AQUEOUS ENVIRONMENTAL SAMPLES

JUDITH SCHENZEL[1,2], RENÉ P. SCHWARZENBACH[1,2], THOMAS D. BUCHELI[1]

[1] AGROSCOPE RECKENHOLZ-TÄNIKON, RESEARCH STATION ART, CH-8046 ZURICH, SWITZERLAND

[2] INSTITUTE OF BIOGEOCHEMISTRY AND POLLUTANT DYNAMICS, SWISS FEDERAL INSTITUTE OF TECHNOLOGY, CH-8092 ZURICH, SWITZERLAND

REPRODUCED WITH PERMISSION FROM:

JOURNAL OF AGRICULTURAL AND FOOD CHEMISTRY, 2010

VOLUME 58, ISSUE 21, PAGES 11207–11217

Abstract

Mycotoxins are naturally occurring secondary metabolites of fungi colonizing agricultural products on the field or during storage. In earlier work, we have shown that two common mycotoxins, i.e. zearalenone and deoxynivalenol, can be present at significant levels in the aquatic environment. This raises the question about the relevance of a wider range of mycotoxins in natural waters. In this investigation we present the first validated method for analysis of some additional 30 mycotoxins in drainage-, river, and waste water treatment plants effluent water. The method includes solid phase extraction over Oasis HLB cartridges, followed by liquid chromatography with electrospray ionization triple quadrupole mass spectrometry. Absolute method recoveries for 13 out of the 33 mycotoxins were higher than 70% in waste water treatment plant effluent (at 25 ng/L) and 27 compounds had MDLs below 10 ng/L. The applicability of this method is illustrated with selected data from our ongoing monitoring campaigns. Specifically, and for the first time, beauvericin and nivalenol were quantified in drainage and river water samples with mean concentrations of 6.7 and 4.3 ng/L, and 6.1 and 5.9 ng/L, respectively. These compounds thus add to the complex mixture of natural and anthropogenic micropollutants in natural waters, where their ecotoxicological risk remains to be evaluated.

2.1. Introduction

Mycotoxins are naturally occurring secondary metabolites of fungi colonizing a variety of cereals, fruits, vegetables, and organic soil material, but can also be produced due to moist conditions during storage. These compounds are therefore commonly found in crop grown and stored for human or animal consumption, as well as in processed food.[1, 2]

More than hundred filamentous fungi are known, e.g., *Alternaria*, *Aspergillus*, *Claviceps*, *Fusarium* and *Penicillium* species (spp.), which produce a plethora of mycotoxins with a great structural diversity, and thus different chemical and physical properties. Among the most prominent substance classes are alternaria toxins, aflatoxins, ergot alkaloids, fumonisins, ochratoxins, resorcyclic acid lactones and trichothecenes. Mycotoxins and their associated health impact on humans and animals have been broadly investigated.[3,4] Several of them (e.g., aflatoxins, ochratoxins, and fumonisins) have been ranked as the most important chronic dietary risk factor, higher than synthetic contaminants, plant toxins, food additives or pesticide residues.[5] Therefore, their occurrence in food and feed has been studied extensively.[6,7]

In contrast, very little is known about the distribution of mycotoxins in the environment and only few studies have been published. Recent work[8-10] demonstrated that the common mycotoxins zearalenone and deoxynivalenol can be emitted into the aquatic environment via drainage and run-off from infested agricultural fields. The concentrations and corresponding emission rates of deoxynivalenol and zearalenone in drainage water were 20-5000 ng/L and 600 mg/ha for deoxynivalenol, and up to 35 ng/L and 3 mg/ha for zearalenone.[10] Comparable amounts of pesticides are emitted from agricultural used fields via drainage water ranging from 3 mg to 56 g per ha.[11] Several authors[12, 13] reported significant correlations between mycotoxin intake and excretions by humans. This indicates that human excretions may be yet another relevant source of mycotoxins in the aquatic environment, depending on the removal rate in waste water treatment plants (WWTP). In fact, Wettstein and Bucheli[14] reported only partial elimination of deoxynivalenol by Swiss WWTP. Together, runoff from agricultural fields and WWTP effluents can lead to the frequent (deoxynivalenol) and occasional (zearalenone) occurrence of mycotoxins in river water.[10] Overall, only fragmentary information on mycotoxin emission pathways and their environmental fate and behavior is available. Compared to deoxynivalenol and zearalenone, other mycotoxins might be produced in even higher amounts in wheat and maize,[15] and have similar or even higher aqueous solubilities. We therefore hypothesize that not only deoxynivalenol and zearalenone, but also a larger number of mycotoxins is likely to enter the aquatic environment primarily via the two pathways indicated above.

The accurate and precise quantification of mycotoxins in the needed low ng/L concentration range in natural waters requires a selective and sensitive analytical method. To achieve such low detection limits, preconcentration by solid phase extraction (SPE) prior to detection is indispensable. The applicability of SPE for the simultaneous quantification of different classes of micropollutants with widely varying chemical and physical properties in various water samples has been demonstrated in earlier publications.[16,17] High performance liquid chromatography electrospray tandem mass spectrometry (HPLC-ESI-MS/MS) has in recent years become the method of choice for separation and detection of mycotoxin in various biological matrices.[7,18] Nonetheless, one drawback of this widely acknowledged technique is signal suppression, which is caused by co-extracted matrix components. Ion suppression has a profound influence on accuracy, precision and sensitivity, and leads to imprecision when quantifying low levels of micropollutants in environmental samples by HPLC-ESI-MS/MS. Hence, there are several methods to compensate for the imprecision caused by ion suppression. Commonly used methods are matrix matched calibration or standard addition, but these methods require an extra effort for analysis in dynamic environmental systems with fast changing matrix composition (e.g., rivers). Despite the fact that ion suppression is a severe problem, it is not always accounted for in the literature.[19,20] The use of isotope labeled internal standards (ILIS) is the most powerful method to overcome matrix effects.[10,14,21] If ILIS are available, their application is a very time-effective alternative to standard addition or matrix matched calibration. Some of the commercially available isotope labeled mycotoxins were used for quantification of mycotoxins in cereals,[22, 23] as well as in environmental trace analysis.[10,14,24]

In this study, we describe an accurate, precise and sensitive analytical method for 33 selected mycotoxins covering the major compound classes in various aqueous samples. The selection of mycotoxins was based on several factors, like frequency of occurrence in food and feed matrices, the chemical and physical properties, as well as the toxicity of the compound. The method comprises SPE followed by separation and detection with HPLC-MS/MS (ESI +/-) and applies seven ILIS for target analytes representing several different compound classes. To our knowledge, this is the first screening method for the quantification of mycotoxins in samples from different aqueous environments. The method is validated for Milli-Q water, drainage water, river water and WWTP effluent, and its application is demonstrated with some initial data from different currently ongoing field studies.

2.2. Materials and Methods

2.2.1 Chemicals

3-Acetyl-deoxynivalenol,15-acetyl-deoxynivalenol, aflatoxin B_1 (≥99%), aflatoxin B_2 (≥98%), aflatoxin G_1 (≥99%), aflatoxin G_2 (≥99%), aflatoxin M_1 (≥99%), citrinin (≥99%), deoxynivalenol (≥99%), HT-2 toxin (≥99%), fumonisin B_1 (≥98%), fumonisin B_2 (≥96%), patulin (≥99%) and T-2 toxin (≥99%) were supplied by Fermentek (Jerusalem, Israel). Alternariol, alternariol monomethylether, altenuene, beauvericin, diacetoxyscirpenol (≥99%), ergocornine, fusarenon-X, neosolaniol, nivalenol, ochratoxin A, ochratoxin B, sterigmatocystin, sulochrin, tentoxin, verrucarin A, zearalenone (≥99%), α-zearalenol (≥98%), and β-zearalenol (≥95%) were supplied by Sigma (Sigma-Aldrich GmbH, Buchs, Switzerland). Fumonisin B_3 and ergocryptine were purchased from Biopure (Referenzsubstanzen GmbH, Tulln, Austria). For structures, see **Figure 2.1.** The six [$^{13}C_x$] ILIS $^{13}C_{17}$-3-acetyl-deoxynivalenol,

$^{13}C_{15}$-deoxynivalenol, $^{13}C_{22}$ HT-2 toxin, $^{13}C_{34}$-fumonisin B_1, $^{13}C_{20}$-ochratoxin A and $^{13}C_{24}$-T-2 toxin were obtained from Biopure (Referenzsubstanzen GmbH, Tulln, Austria). All compounds were delivered diluted in acetonitrile (MeCN), except $^{13}C_{34}$-fumonisin B_1 which was diluted in a MeCN/Milli-Q water (1/1, v/v) mixture. Concentrations of the ILIS were 25 µg/mL, despite ochratoxin A with 10 µg/mL. D_6-zearalenone was prepared in our own laboratory as described earlier.[24]

Methanol (MeOH, HPLC-grade) and MeCN (HPLC-grade) were purchased from Scharlau (Barcelona, Spain). Ammonium acetate (p.a.) and acetic acid (p.a.) were supplied by Fluka (Fluka AG, Buchs, Switzerland). Deionised water was further cleaned with a Milli-Q gradient A10 water purification system from Millipore (Volketswil, Switzerland). High-purity N_2 (99.99995%) and Ar (99.99999%) were obtained from PanGas (Dagmarsellen, Switzerland).

Mycotoxin crystalline compounds were dissolved in pure MeCN, except fusarenon-X, neosolaniol, ochratoxin A, and ochratoxin B, which were dissolved in an MeCN/Milli-Q water (1/1, v/v) mixture. The individual stock solutions contained concentrations ranging from 100 to 1000 mg/L. Multicomponent stock solutions were prepared in MeOH in concentrations of 100, 500 and 1000 ng/mL. The internal standard solutions were diluted with MeOH to obtain a concentration of 1 mg/L, except D_6-zearalenone with a concentration of 2 mg/L. All compounds, either in single-component or multicomponent MeOH stock solution, were stored at +4°C in the dark. Aqueous calibration standards holding all mycotoxins equivalent to the concentration range of 1 – 100 ng/L, and all seven ILIS at 50 ng/L, were freshly prepared in Milli-Q water/MeOH (9/1, v/v) from the multicomponent stock solutions each time a new series of samples was analyzed.

Structures of aflatoxins [AFs] including Aflatoxin B_1, Aflatoxin B_2, Aflatoxin M_1, Aflatoxin G_1, Aflatoxin G_2

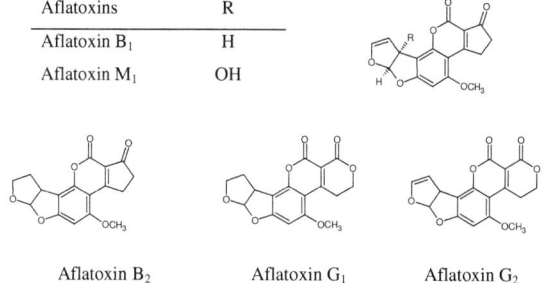

Aflatoxins	R
Aflatoxin B_1	H
Aflatoxin M_1	OH

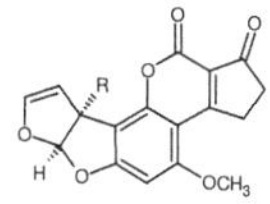

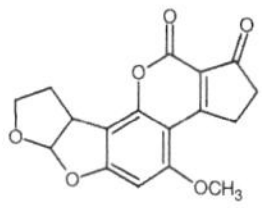

Aflatoxin B_2

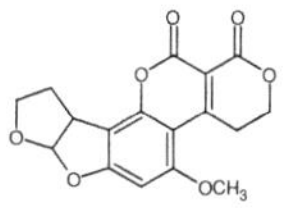

Aflatoxin G_1

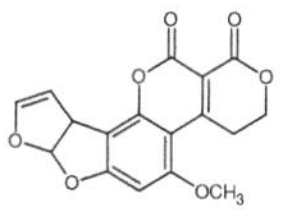

Aflatoxin G_2

Structures of Alternaria toxins:

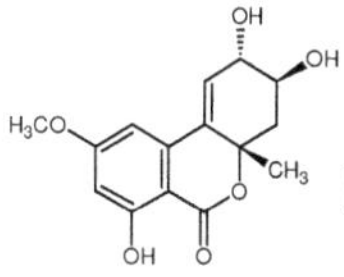

Altenuene

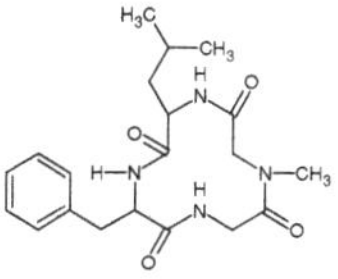

Tentoxin

Alternariol R=OH

Alternariolmethylether R=OCH_3

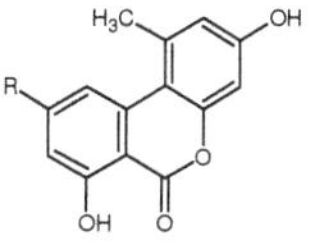

Alternariol

Structures of ergot alkaloids:

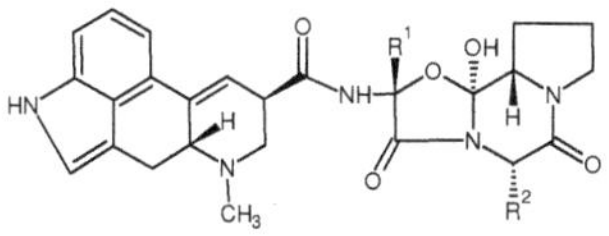

Analyte	R^1	R^2
Ergocornine	$CH(CH_3)_2$	$CH(CH_3)_2$
Ergocryptine	$CH(CH_3)_2$	$CH_2CH(CH_3)_2$

Structures of fumonisins:

Analyte	R^1	R^2
Fumonisin B_1	OH	OH
Fumonisin B_2	OH	H
Fumonisin B_3	H	OH

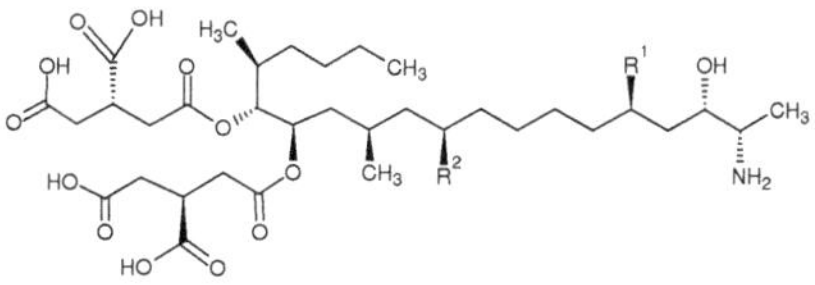

Structures of ochratoxins:

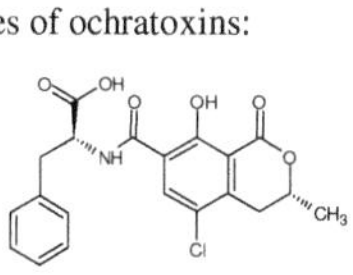

Ochratoxin A

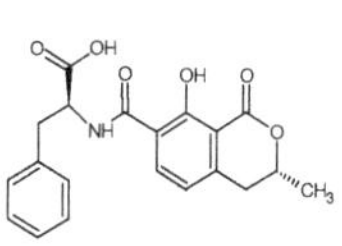

Ochratoxin B

Figure 2.1. Chemical structures of the investigated compounds.

Structures of "Other" toxins:

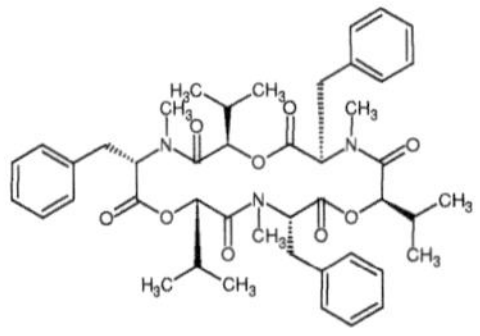

Beauvericin

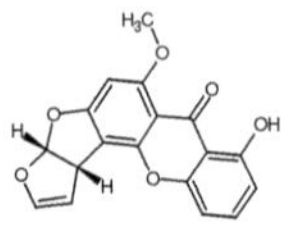

Sterigmatocystin

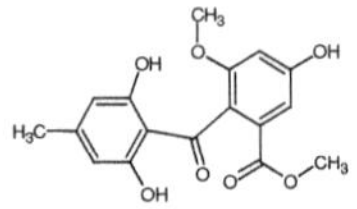

Sulochrin

Structures of Penicillium toxins:

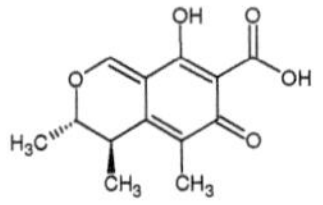

Citrinin

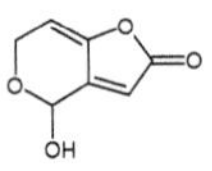

Patulin

Structures of trichothecenes:

Trichothecene	R^1	R^2	R^3	R^4	R^5
Type A					
Diacetoxyscirpenol	OH	$OCOCH_3$	$OCOCH_3$	H	H
HT-2 toxin	OH	OH	$OCOCH_3$	H	$OCOCH_2CH(CH_3)_2$
Neosolaniol	OH	$OCOCH_3$	$OCOCH_3$	H	OH
T-2 toxin	OH	$OCOCH_3$	$OCOCH_3$	H	$OCOCH_2CH(CH_3)_2$
Type B					
Deoxynivalenol	OH	H	OH	OH	=O
Fusarenon-X	OH	$OCOCH_3$	OH	OH	=O
Nivalenol	OH	OH	OH	OH	=O
3-AcDON	$OCOCH_3$	H	OH	OH	=O
15-AcDON	OH	H	$OCOCH_3$	OH	=O

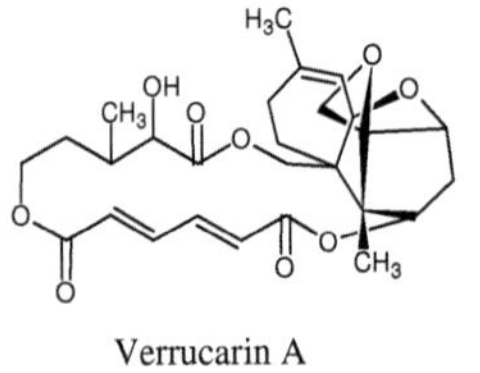

Verrucarin A

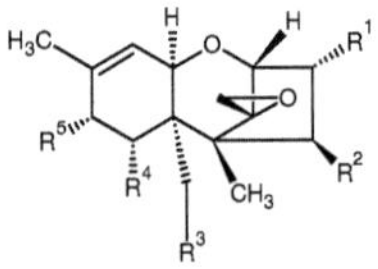

Trichothecene

Structures of resorcyclic acid lactones:

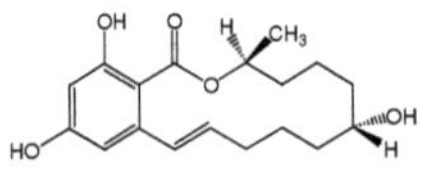

α-Zearalenol

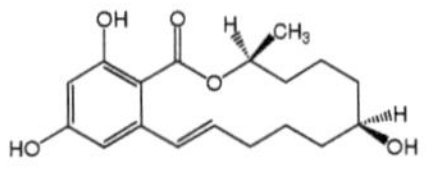

β-Zearalenol

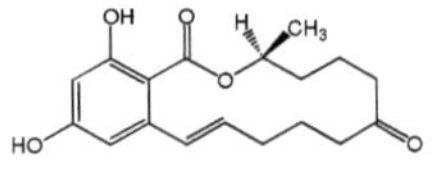

Zearalenone

Figure 2.1. continued.

2.2.2 Sample Collection and preparation

Drainage water samples were collected at our field study site at Reckenholz, Switzerland[8] using portable automatic flow proportional samplers (Teledyne Isco Inc., Lincoln, NE, USA). Surface water samples were obtained weekly and fortnightly from the monitoring program from the Canton of Zurich (Office for Waste, Water, Energy, and Air; AWEL), and the monitoring program of the Swiss government (National Long-Term Surveillance of Swiss Rivers; NADUF), respectively. Details have been published previously.[25] Grab samples of WWTP effluents were collected at the Kloten/Opfikon (Zurich, Switzerland) facility.[14]

Raw water samples were filtered (glass fibre filters, pore size 1.2 µm, Millipore, Volketswil, Switzerland) by vacuum filtration (Supelco, Bellfonte PA, USA), transferred to 1 L glass bottles and stored in the dark at +4 °C until analysis within 2 weeks (storage tests showed that mycotoxins were stable over this period of time). Before SPE, the pH was adjusted to between 6.6 and 7.0 by adding either ammonium acetate or acetic acid. In routine analysis the exact volume of 1 L was spiked with 50 µL of the ILIS mixture before the storage or processing of the sample. During method validation the ILIS was spiked as stated under the method validation parameters. The samples were shaken vigorously before further treatment.

2.2.3 Solid Phase Extraction (SPE)

Filtered water samples (1 L) were concentrated and purified by performing reversed-phase SPE (Oasis HLB cartridges, 6 mL, 200 mg; Waters Corporation, Milford MA, USA) on a 12-fold vacuum extraction box (Supelco, Bellfonte PA, USA). The SPE cartridges were consecutively conditioned with 5 mL of MeOH, 5 mL of Milli-Q water and MeOH (1/1, v/v), and 5 mL of Milli-Q water. One Liter water samples were passed through the cartridges with a maximum flow rate of 10 mL/min. Subsequently, the cartridges were washed with 5 mL of Milli-Q water. Without any additional column drying step the analytes were eluted with 5 mL MeOH and the aliquots were collected in conical reaction vial vessels (Supelco, Bellfonte PA, USA). The 5 mL MeOH aliquots were reduced to 100 µL under a gentle nitrogen gas stream at 50°C. The extracts were reconstituted in 900 µL of Milli-Q water/MeOH (90/10, v/v) and transferred into 1.5 mL amber glass vials. The samples were stored at +4°C and analyzed within 48 h.

The following other SPE phases were initially tested according to the manufacturer's guidelines for their mycotoxin extraction efficiency: Bond Elut Plexa cartridges 6 mL, 200 mg (VarianInc. Lake Forest CA, USA); Chromabond HR-X cartridges 6 mL, 200 mg (Macherey-Nagel GmbH & Co. KG, Düren, Germany).

2.2.4 LC-MS/MS Analysis

LC-MS/MS was performed on a Varian 1200L LC-MS instrument (VarianInc, Walnut Creek CA, USA). The mycotoxins were separated using a 125 mm × 2.0 mm i.d., 3 µm, Pyramid C_{18} column, with a 2.1 mm x 20 mm i.d. 3 µm guard column of the same material (Macherey-Nagel GmbH & Co. KG, Düren, Germany) at room temperature. Two different chromatographic runs were used for the separation of all compounds, on in the positive and one in the negative ionization mode, respectively. The optimized LC mobile phase gradient for the analysis of the analytes measured in negative ionization mode was as follows: 0.0 min: 0% B (100% A), 2.0 min: 0% B, 15.0 min: 100% B, 18.0 min: 100% B, 19.0 min: 0% B, 24.0 min: 0% B. The analytes measured in the positive ionization mode were separated with the following gradient: 0.0 min: 27% B (73% A), 1.0 min: 27% B, 1.3 min: 45% B, 5.0 min: 45% B, 5.3 min: 63% B, 9.0 min: 63% B, 9.3 min: 81% B, 13.0 min: 81% B, 13.3 min: 100% B, 20.0 min: 100% B, 21.0 min: 27% B, 24.0 min: 27% B. In both cases eluent A consisted of Milli-Q water/MeOH/acetic acid (89/10/1, v/v/v) and eluent B of Milli-Q water/MeOH/acetic acid (2/97/1, v/v/v). Both eluents were buffered with 5 mM ammonium acetate. The injection volume was 40 µL and the mobile phase flow was 0.2 mL/min.

LC-MS interface conditions for the ionization of the acidic mycotoxins in the –ESI mode were: needle voltage - 4000 V, nebulizing gas (compressed air) 3.01 bar, drying gas (N_2, 99.5%) 275°C and 1.24 bar, shield voltage -600 V. The neutral mycotoxins were ionized in the +ESI mode: needle voltage + 4500 V, nebulizing gas (compressed air) 3.01 bar, drying gas (N_2, 99.5%) 275°C and 1.24 bar, shield voltage + 600 V.

The analyte-dependent MS and MS/MS parameters, and collision cell energies were acquired by direct infusion of standards with concentrations of 4 ppm into the MS using an external syringe pump (Pump 11, Harvard Apparatus, Holliston MA, USA) with a flow rate of 50 µL/min. The main fragments were identified using multiple reaction monitoring (MRM). Precursor-product ion transitions and corresponding collision cell voltages were obtained for each analyte. A second product ion was monitored for identity confirmation. The collision cell gas (Ar, 99.999%) pressure was 2.0(±0.1)e^{-6} Torr and the detector voltage was set to 1800V. Depending on the number of ion transitions per time segment, scan times ranged from 0.1 to 0.4, and 0.2 to 2.4 s for acidic and neutral compounds, respectively. Scan widths were 1.0 *m/z* for Q1 and 1.5 *m/z* for Q3, and mass peak width was 0.7 *m/z*.

Analytes without ILIS were quantified using matrix matched calibrations. If not stated otherwise, the analytes with corresponding ILIS were quantified using the internal standard method, i.e., calibration standards in Milli-Q water containing increasing amounts of analytes and constant amounts of the seven ILIS (see above) and matrix matched calibrations. Data processing was carried out using the Varian MS Workstation (v. 6.9.2.) software.

2.2.5 Method validation parameter

Linearity: The linearity of the MS/MS detector was tested with Milli-Q water/MeOH (9/1, v/v) containing mycotoxins at concentrations between 0.1 ng/L and 100 µg/l, corresponding to 0.005 – 5 ng at the detector, respectively.

The linearity for the three different environmental matrices was tested for matrix calibrations ranging from 1 to 100 ng/L.

Ion suppression: Matrix effects during analyte ionization causing suppression or enhancement of the analyte signal were evaluated by comparing the obtained signal from injection of the same amount of analyte in Milli-Q water/MeOH (9/1, v/v) and an extracted matrix blank. For this purpose, 8 L of each matrix (Milli-Q, drainage, river, and WWTP effluent water) were filtered and concentrated on an individual SPE-column following the procedure as described above. The eight obtained SPE eluates were combined and afterwards divided into eight equal portions. Matrix matched calibration samples were obtained by carrying out standard addition to the final extracts to produce concentration equivalent to those of 1 L samples containing 1, 5, 10, 25, 50 and 100 ng of each analyte. Additionally, 50 ng of ILIS were added to each portion. Calibration curves were obtained by plotting measured analyte peak areas against corresponding analyte concentration levels in pure solvent and in the extracted matrix, respectively. Linear regression was performed for each curve. The ion suppression (expressed in percentage) was quantified as 1 minus the ratio between the slope of the curve obtained for the final extracts and the slope of the curve for the pure solvent (see **SI**).

Absolute and relative SPE recoveries, absolute and relative method recoveries: Absolute SPE recoveries were determined for all mycotoxins in Milli-Q water testing three different extraction cartridge materials. Therefore, 1 L samples were spiked with mycotoxins prior to SPE to produce concentration levels of 10, 25, 50 and 100 ng/L. ILIS (50 μL) was added directly into the 5 mL eluates extracted from the cartridge. The matrix matched SPE calibration curve was obtained by carrying out standard addition to the pure Milli-Q matrix SPE eluates. The absolute SPE recovery for the analytes without an ILIS was defined as the ratio of the slope obtained by adding the analyte before the extraction to the slope of the matrix matched SPE calibration (see SI). The absolute SPE recovery for the compounds with corresponding ILIS was defined as the ratio of the slope resulting from the measured area of the analyte added before the extraction to the measured area of the corresponding ILIS added to the SPE eluate, divided by the slope of the matrix matched SPE calibration (see **SI**).

Absolute method recoveries were determined for all mycotoxins in Milli-Q, drainage, river, and WWTP effluent water. One liter samples were spiked with mycotoxins prior to SPE to produce concentration levels of 5, 25 and 100 ng/L. Five replicates were prepared for each concentration level. One liter of native matrix was tested for native mycotoxin contents and/or contamination due to the addition of ILIS. After SPE and eluate evaporation, sample residues were reconstituted in 650 μL Milli-Q water and 50 μL of each ILIS solution was added. The matrix matched calibration curve for the determination of absolute and relative recovery was obtained by carrying out standard addition to the final extracts to produce concentration equivalent to those of 1 L samples containing 1, 5, 10, 25, 50 and 100 ng of each analyte. The absolute method recovery for the analytes without an ILIS was defined as the ratio of the slope obtained by adding the analyte before the extraction to the slope of the matrix matched calibration (for details see **SI**). The absolute method recovery for the analytes for which an ILIS was available was defined as the ratio of the slope resulting from the measured area of the analyte added before the extraction to the measured area of the corresponding ILIS added after the extraction, divided by the slope of the matrix matched calibration (see **SI**).

The relative method recovery was determined only for those analytes which have an ILIS. It was defined as the ratio of the slope of the measured area of analyte and ILIS added before the extraction, divided by the slope of the matrix matched calibration (see **SI**).

Method and instrument precision, method and instrument detection limits: The method precision (MP) values presented in the results and discussion part were determined as the relative standard deviation (RSD) of five replicates at the concentration level of 5 ng/L, except for those compounds exhibiting higher MDL. The instrument precision (IP) was obtained from five consecutive analyses of an individual sample.

The method detection limit (MDL) was defined as three times the absolute standard deviation of the five replicates at the lowest quantifiable concentration level.[26] The instrument detection limit (IDL) was defined as three times the absolute standard deviation of five standard calibration replicates at the lowest concentration level. For each sample type, one liter of unfortified matrix, i.e. containing ILIS only, was analyzed for its native mycotoxin content and/or contamination due to the addition of ILIS. No mycotoxins were quantified in any of these control samples.

2.3. Results and Discussion

2.3.1 Optimization of the extraction method

The mycotoxin extraction procedure was optimized with respect to their recoveries over different SPE cartridges, elution solvent and pH value of the aqueous sample. For the extraction of multiple components covering a broad range of polarities, and acid/base properties, usually polymeric, end-capped functionalized adsorber materials are applied.[27] Two such materials, i.e., Chromabond HR-X and Bond Elut Plexa cartridges, were tested of their enrichment efficiency. The OASIS HLB was additionally chosen due to its wide application range and the ability to effectively retain a large number of compounds, basic, neutral and acidic.[16,28] It is designed, in fact, to retain both hydrophilic and hydrophobic compounds with high capacity by means of both van der Waals and H-donor–H-acceptor interactions.

With a median SPE recovery of 95%, Oasis HLB exhibited the best performance for analytes measured in the negative ionization mode (see SI). Only two compounds, alternariol monomethylether and nivalenol, showed recoveries below 30%. According to good laboratory practice, recoveries ranging from 70-120% are excellent recoveries.[29] The other two tested cartridge materials showed median recoveries below 90% in the negative ionization mode. Bond Elut Plexa exhibited the best cartridge recoveries for the analytes measured in positive ionization mode (see **SI**), while Oasis HLB had a median recovery of

90%. Consequently, Oasis HLB was chosen as extraction cartridge. Because fumonisins were not eluted when using acetonitrile as extracting solvent,[30] and due to the supplier's advice, 5 mL of methanol was chosen as mycotoxin eluent from the cartridge material.

No significant differences in absolute SPE recoveries were observed between pH 5 and 7. Given the fact that many surface water samples have a pH close to 8, the sample pH was further on adjusted with acetic acid to lie in between 6.6 and 7.0.

2.3.2 Chromatographic separation and MS detection

The reversed phase liquid chromatography separation of the mycotoxins was optimized with respect to LC column type, mobile phase, and gradient dynamics to achieve the greatest possible selectivity and sensitivity. The Nucleodur Gravity reversed phase C_{18} column was chosen as the stationary phase because well-resolved peak shapes were generally obtained for all analytes despite their chemical diversity. The often difficult separation of ergot alkaloid epimers was improved by adding 1% of acetic acid into the eluent solutions.[31] Quantification problems due to intensity differences of the MRM transition of the two epimers were overcome by building the sum of the two epimer peaks. As a representative example, a chromatogram of a WWTP effluent extract spiked with 25 ng/L of each investigated compound is shown in the **SI**.

The mycotoxins were either ionized to their $[M\text{-}H]^-$ or $[M\text{+}H]^+$ form, respectively, or to corresponding ammonia- and acetate adducts (**Table 2.1**). The polarity showing the more abundant precursor ions was selected for each analyte. Thirteen compounds were optimized using the negative ionization mode, because several of them, like the alternaria toxins exhibited no detectable MS signal in the positive ionization mode. Twenty compounds were measured in the positive ionization mode, like the aflatoxins, which exhibited no detectable MS signal in the negative ionization mode. Due to the large number of analytes, polarity switching during the chromatographic run was not possible. Therefore, both positive and negative ESI modes were used in two separate chromatographic runs to assure stable and sensitive MS conditions for all analytes. Due to the addition of 5 mM of ammonium acetate to the chromatographic eluents, some analytes (A-trichothecenes) formed $[M+NH_4]^+$ adducts with higher precursor signal intensities than the related $[M+H]^+$ species. The monitored product ions from these adduct precursors were analyte specific, and selected to fulfill the requirement for the confirmation of substances according to the Annex I of Directive 96/23/EC.[32] The parameters of the optimized MS detection, including capillary and collision cell voltages, and MRM transitions of each analyte are summarized in **Table 2.1**.

Table 2.1. Analytical parameters for the selected mycotoxins and their available corresponding isotope labeled internal standards (ILIS).

Compound	Retentio n time	Precurs or ion	Adduct	Main product ion		Secondary product ion		Capillary Voltage
					Collisi on energy		Collisi on energy	
		[*m/z*]		[*m/z*]	[eV]	[*m/z*]	[eV]	[V]
Aflatoxins								
Aflatoxin B_1	7.52	313	$[M+H]^+$	285	20	270	30	65
Aflatoxin B_2	6.37	315	$[M+H]^+$	287.1	20	259.1	30	65
Aflatoxin G_1	5.44	329	$[M+H]^+$	243	28	214.7	30	60
Aflatoxin G_2	4.68	331	$[M+H]^+$	313.1	20	245.1	30	65
Aflatoxin M_1	4.61	329	$[M+H]^+$	273	20	259	20	65
Alternaria toxins								
Altenuene	14.52	291	$[M-H]^-$	247.9	20	275.8	20	60
Alternariol	19.43	271.1	$[M-H]^-$	255.7	20	227.6	20	60
Alternariol	14.61	257	$[M-H]^-$	212.8	20	214.8	20	70
Tentoxin	16.34	413.2	$[M-H]^-$	271	10	140.9	10	65
Ergot alkaloids								
Ergocornine	8.19/10.9	562.3	$[M+H]^+$	544	10	305.2	30	40
Ergocryptine	9.57/12.3	576.3	$[M+H]^+$	558.4	10	305.2	30	50
Fumonisins								
Fumonisin B_1	12.02	722.3	$[M+H]^+$	352.3	30	334.4	50	80
$^{13}C_{34}$ Fumonisin	*11.97*	*756.8*	*$[M+H]^+$*	*374.7*	*30*	*738.8*	*13*	*80*
Fumonisin B_{2+3}	14.57	706.3	$[M+H]^+$	336.4	40	318.4	40	75
Ochratoxin A	13.63	404.2	$[M+H]^+$	239.2	20	358.1	10	40
$^{13}C_{20}$ Ochratoxin	*13.59*	*424.4*	*$[M+H]^+$*	*250*	*20*	*377.3*	*10*	*40*
Ochratoxin B	11.56	370.2	$[M+H]^+$	205	20	324.1	11	40
Others								
Beauvericin	19.77	784.4	$[M+H]^+$	244	20	262.4	26	70
Sterigmatocystin	14.98	325	$[M+H]^+$	310.1	20	281.3	40	35
Sulochrin	6.89	333.2	$[M+H]^+$	208.9	10	136.2	40	25
Penicillium								
Citrinin	16.24	281.1	$[M+Ac]^-$	204.8	20	248	20	35
Patulin	2.97	152.9	$[M-H]^-$	108.9	10	80.9	10	30
Resorcyclic acid lactones								
α-Zearalenol	17.88	319.4	$[M-H]^-$	275	20	160	30	60
β-Zearalenol	16.81	319.1	$[M-H]^-$	275.2	20	301.1	20	60
Zearalenone	18.23	317.4	$[M-H]^-$	175	20	131	25	60
D_6-Zearalenone	*18.19*	*323.4*	*$[M-H]^-$*	*134*	*30*	*279*	*20*	*60*
Trichothecene A								
Diacetoxyscirpe	7.18	384.3	$[M+NH_4$	307	10	247	12	30
HT-2 toxin	10.32	442.2	$[M+NH_4$	263.1	12	215.2	12	30
$^{13}C_{22}$ HT-2 toxin	*10.32*	*464.4*	*$[M+NH_4$*	*278.2*	*12*	*215.8*	*12*	*30*
Neosolaniol	3.05	400.2	$[M+NH_4$	305	11	185	20	20
T-2 toxin	12.43	484.4	$[M+NH_4$	185.1	21	215	17	35
$^{13}C_{24}$ T-2 toxin	*12.43*	*508.6*	*$[M+NH_4$*	*198*	*17*	*229.4*	*13*	*35*
Verrucarin A	12.44	520.1	$[M+NH_4$	249.1	20	231.2	20	45
Trichothecene B								
3-Acetyl-DON	11.76	397.1	$[M+Ac]^-$	307.1	18	337.1	14	25
$^{13}C_{17}$ 3-AcDON	*11.77*	*413.5*	*$[M+Ac]^-$*	*323.6*	*14*	*182.9*	*30*	*25*
15-Acetyl-DON	4.14	339.2	$[M+H]^+$	321.6	20	137	30	35
Deoxynivalenol	5.06	355.1	$[M+Ac]^-$	295.4	10	264.8	6	50
$^{13}C_{16}$ DON	*5.01*	*370*	*$[M+Ac]^-$*	*310*	*10*	*279*	*10*	*50*
Fusarenone-X	9.19	413.1	$[M+Ac]^-$	353	14	263	11	35
Nivalenol	2.66	371.1	$[M+Ac]^-$	280.9	10	311	6	35

2.3.3 Validation of the optimized method

Linearity: The MS/MS was linear for matrix solutions between 0.1 ng/L and 100 µg/L, corresponding to 5 pg to 5 µg at the detector, respectively (0.9410 < R^2 < 0.9998). The linearity was also tested with standard solutions ranging up to concentrations of 500 µg/L. For aflatoxin B_2, alternariol, alternariol monomethylether, citrinin, ergocornine, T-2 toxin, tentoxin, and zearalenone non-linear conditions were reached in the concentration range of 400 to 500 µg/L. The varying aqueous matrices did not show any effects on the MS/MS linearity.

Ion suppression: Ion suppression was quantified for all analytes in all investigated matrices and reached up to 94% (**Figure 2.2**). The negative values for some analytes of the class of trichothecenes A and B correspond to a signal enhancement caused by matrix interference. Generally, the extent of ion suppression depends on the type of matrix, as shown in **Figure 2.2 A** and **B**. Overall, ion suppression was consistently higher in WWTP effluents than in drainage and in river water. This is plausible considering the high output of organic material from WWTPs. As expected, lowest, but sometimes still considerable ion suppression was quantified in Milli-Q water. Reasons for this could be exogenous substances from external sources during sample preparation, for example residues released from the SPE cartridge or buffers.[33] Generally, it was observed that ion suppression of the analytes was more pronounced in ESI- than ESI+ mode (**Figure 2.2**), which is in agreement with earlier investigations.[34] The explicit ion suppression data quantified for all analytes in each matrix is listed in the SI.

Unfortunately, there are only four publications available with ion suppression data for mycotoxins in comparable matrices.[10,14,24,35] Lagana et al.[35] reported APCI- ion suppression of 15-25% for different resorcyclic acid lactones in aqueous environmental samples, and Wettstein and Bucheli[14] reported 18%, 14%, and -20% APCI- ion suppression for deoxynivalenol in drainage, river, and WWTP effluent water, respectively. These numbers contrast to the ESI- ion enhancement observed in the present study (-14%, -8%, and -61% for drainage, river, and WWTP effluent water, respectively, **SI**), and may be explained with the different ionization methods in use. Hartmann et al.[24] reported ESI- ion suppression values of 33-57% in drainage, 28-54% in river, and 56-68 % in WWTP effluent waters for resorcyclic acid lactones. These are somewhat lower than the data reported here (drainage: 12-80%, river: 22-86%, WWTP effluent: 57-88% see **SI**).

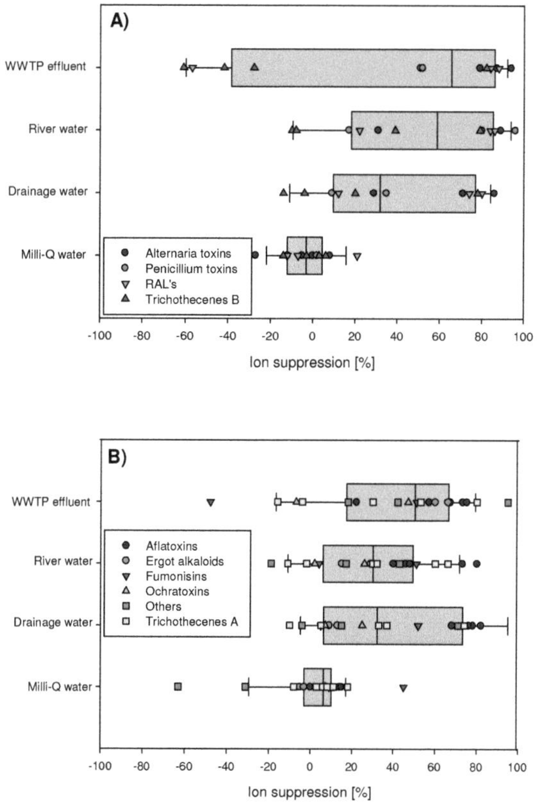

Figure 2.2. Ion suppression of A: mycotoxins measured in negative ionization mode (*-ESI*) and B: mycotoxins measured in positive ionization mode (*+ESI*). Line in box: median (50th percentile); box margins: 25th and 75th percentile; lines with whiskers: 10th and 90th percentile.

Absolute method recoveries: The absolute recoveries were determined for all selected mycotoxins in MilliQ water, drainage water, river water, and WWTP effluent at three different concentration levels (5, 25, and 100 ng/L). All data are presented in the SI, and the results obtained with 100 ng/L are visualized in **Figure 2.3**. In MilliQ water, the absolute recoveries for all 33 mycotoxins ranged from 20 to 109% at 100 ng/L (median 76%). These recoveries are lower than the SPE recoveries presented in **SI**, which is probably due to analyte losses during SPE eluate evaporation and reconstitution.

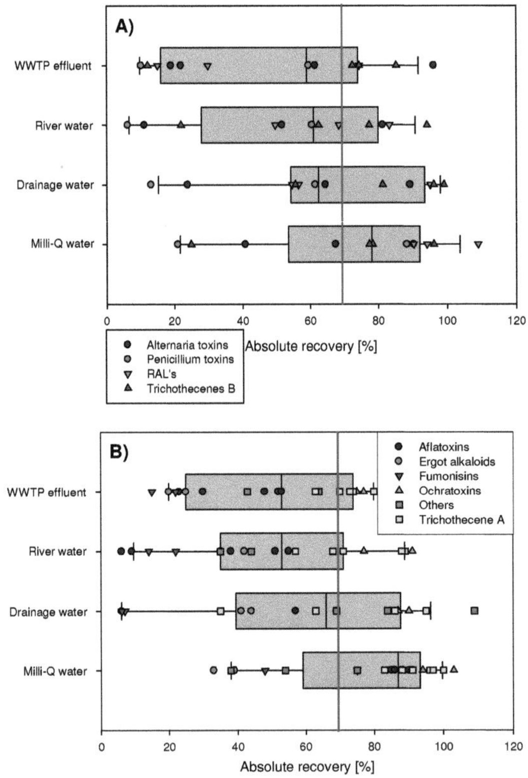

Figure 2.3. Absolute method recovery of A: mycotoxins measured in negative ionisation mode (*-ESI*) and B: mycotoxins measured in positive ionisation mode (*+ESI*) at the 100 ng/L concentration level in various natural waters. The red line indicates the lower end of the satisfactory recovery range (70%). Line in box: median (50th percentile); box margins: 25th and 75th percentile; lines with whiskers: 10th and 90th percentile.

At a concentration level of 25 ng/L, the absolute recoveries for all 33 mycotoxins were in the range of 17 to 102% (median 70%). At the 5 ng/L concentration level, 30 out of 33 compounds could be quantitated and showed absolute recoveries of 16 to 104%. Median absolute recoveries at 5 ng/L were 77%. The natural samples took their toll of the absolute recoveries, and increasingly so for drainage, river and WWTP effluent waters, i.e., matrices of increasing complex nature in terms of dissolved organic carbon content or ionic strength. For instance, at the 100 ng/L spike level, median absolute recoveries dropped from 76% in MilliQ water to 62%, 56%, and 57% in drainage, river, and WWTP effluent water, respectively. Simultaneously, the number of compounds with satisfactory absolute recoveries above 70% was reduced from 23 in MilliQ water to 12, 10, and 13 in drainage, river, and WWTP effluent water, respectively. This trend was corroborated by, and even more pronounced for the lower spike levels (see **SI**).

Specifically, compounds with absolute recoveries below 50% (or spike concentrations below detection limits) were 15-acetyl-deoxynivalenol, the three aflatoxins G_1, G_2, and M_1), alternariol monomethylether, alternariol, the two ergot alkaloids (ergocornine, ergocyrptine), the two fumonisins (B_1, B_{2+3}), nivalenol, patulin, sulochrin α- and β-zearalenol, whereas aflatoxin B_1 and aflatoxin B_2, citrinin, neosolaniol, sterigmatocystin, and tentoxin exhibited values between 50% and 70%. The low absolute recoveries of some of these compounds can be explained by their acidity (e.g., nivalenol, patulin), resulting in very low extraction efficiencies (see **SI**). If quantifiable, recoveries above 70% were predominantly found for 3-acetyl-deoxynivalenol, altenuene, beauvericin, diacetoxyscirpenol, deoxynivalenol, fusarenon-X, HT-2, ochratoxin A and B, T-2, verrucarin A, and zearalenone.

Whereas the here presented absolute recoveries of deoxynivalenol in drainage (99%) and river water (91-104%) were above those reported by Bucheli et al.[10] (85% and 87%, respectively), they were lower (33-72%) than those obtained by Wettstein & Bucheli[14] in WWTP effluent (91%). Lagana et al.[35] reported absolute method recoveries of 85-92% for different resorcyclic acid lactones in aqueous environmental samples, whereas Hartmann et al.[24] reported values of 73-82% in Milli-Q water, 84-103% in drainage water, 92-98% in river water, and 93-94% in WWTP effluents, respectively. These are somewhat higher than the data reported here (drainage: 54-104%, river: 49-95%, WWTP effluent: 14-80%; see SI). Overall, it is in the nature of multi residue methods that one has to cope with a broad range of absolute recoveries. For instance, various classes of concomitantly analyzed pharmaceuticals were recovered from river water at 17 to 101%[36], 6 to 98% absolute method recoveries were reported for 26 biocides, 5 UV filters and 5 benzothiazoles in WWTP and surface waters.[17]

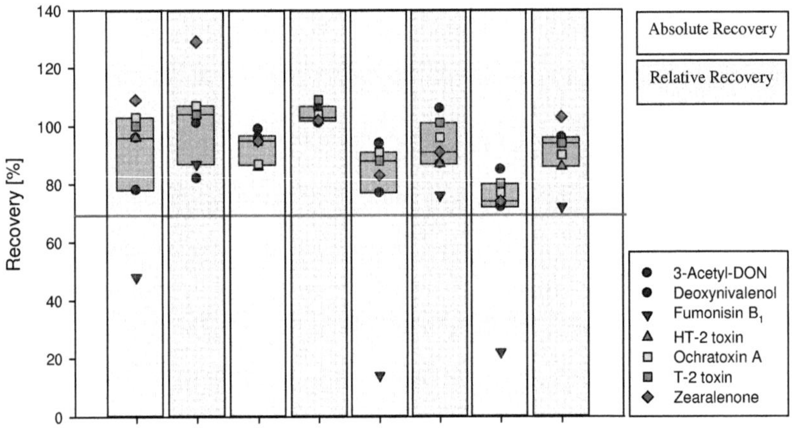

Figure 2.4. Absolute and relative recoveries of mycotoxins with available ILIS at a concentration of 100 ng/L in various natural waters. The red line indicates the lower end of the satisfactory recovery range (70%). Line in box: median (50th percentile); box margins: 25th and 75th percentile.

Relative method recoveries: If available, ILIS were applied to compensate for the variability of matrix effects which affect SPE as well as analyte ionization. The effectiveness of ILIS is illustrated by the comparison of absolute and relative recoveries of seven mycotoxins in various aqueous matrices (**Figure 2.4**). Their absolute recoveries of 49-109%, 86-99%, 14-94%, and 22-85% in MilliQ, drainage, river and WWTP effluent water, respectively, translated into relative recoveries of 82-129%, 101-109%, 76-106%, and 72-103% in these waters at the 100 ng/L spike level. Similarly enhanced recoveries were observed at the lower spike levels of 5 and 25 ng/L (see **SI**). Hence, even in the case of some critical compounds with unsatisfactory absolute recoveries such as fumonisin B_1, the application of a corresponding ILIS provides dependable relative recoveries, allowing for a quantitative analysis of even such problematic analytes.

Precision (method and instrument), blank levels and method detection limits, repeatability: The values of the overall MP ranged from 2 to 28% (median: 4%), 4 to 17% (median: 4%), 2 to 37% (median: 4%), and 2 to 19% (median: 4%) in MilliQ, drainage, river, and WWTP effluent water (see **SI**). Considering the structural diversity of the analytes, the precision obtained for this multi component method are satisfactory,[31, 36] and confirm the robustness of the analytical method. In comparison, the IP was between 1 and 21% for all mycotoxins in all matrices. Specifically, the median number was 4%, 5%, 6%, and 7% in MilliQ, drainage, river and WWTP effluent water (for details see **SI**). This indicates that most of the uncertainty is caused by sample analysis and not sample preparation.

The MDL ranged from 0.2 to 5.2 ng/L (median: 0.8 ng/L), 0.3 to 44.9 ng/L (median: 0.9 ng/L), 0.3 to 29 ng/L (median: 1 ng/L), and 0.4 to 47.7 ng/L (median: 1.0 ng/L) in MilliQ, drainage, river and WWTP effluent water (**Figure 2.5 A** and **B**; for details see **SI**). The IDL ranged from 0.1 to 5.1 ng/L (median: 0.3 ng/L) obtained for standard calibration solutions. The MDLs obtained here can be compared with earlier methods specializing on the quantification of individual target analytes in natural waters. Hartmann et al.[24] reported MDL of 0.9-1.4 ng/L, 0.5-2.1 ng/L, 0.6-1.1 ng/L, and 0.8-12.4 ng/L for resorcyclic acid lactones in MilliQ, drainage, river and WWTP effluents, respectively. Whereas the data presented here was similar in MilliQ water (0.8 – 1.1 ng/L), it proved inferior in natural waters (6.0-44.9 ng/L, 4.9-18.6 ng/L, and 12.9-31.1 ng/L in drainage, river, and WWPT effluent water, respectively). Once more, this discrepancy is a result of accommodating various chemically diverse target analytes in one multi residue method. However, it performed equally well in the case of deoxynivalenol (MDL: 0.8-1.3 ng/L) as previously reported[14] (MDL: 0.8-1.5 ng/L). Generally, the MDL obtained here are comparable to those published for multiple classes of acidic and neutral pharmaceuticals and personal care products,[16,36] or even better in comparison to the application of a multi mycotoxin method for the analysis in beer.[30]

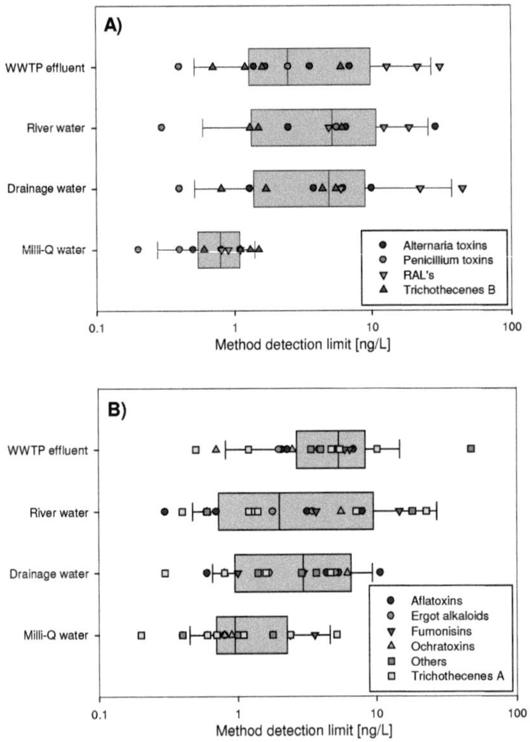

Figure 2.5. Method detection limits (MDL) of A: mycotoxins measured in negative ionization mode (-*ESI*) and B: mycotoxins measured in positive ionization mode (+*ESI*) at low concentrations in various natural waters. Line in box: median (50th percentile); box margins: 25th and 75th percentile; lines with whiskers: 10th and 90th percentile.

2.3.4 Environmental application

The validated analytical multi residue method presented here is currently applied to verify the presence of mycotoxins in natural waters. Firstly, the method is used to study the emission of mycotoxins from a drained test field cultivated with winter wheat at Reckenholz, Zurich (Switzerland).[8] The wheat was artificially infected with four different mycotoxin producing *Fusarium* species, specifically *F. graminearum*, *F. crockwellense*, *F. poae* and *F. avenaceum*, in June 2009 and 2010. Flow-proportional drainage water samples have been collected regularly prior to and following this infection. Initial results of this still ongoing field study are summarized in **Table 2.2**. Deoxynivalenol was detected in 22 out of 129 drainage water samples, with a maximum concentration of 22.5 ng/L. Such numbers are about 1000 times lower than during summer,[10] due to pre-infection and early vegetation conditions during the sampling period shown here. Whereas nivalenol was detected at about the same frequency as deoxynivalenol, beauvericin was even more prevalent. Mean concentrations of nivalenol and beauvericin were 6.7 ng/L and 6.1 ng/L, respectively.

Secondly, seasonal WWTP effluent samples were collected at the Kloten/Opfikon (Zurich, Switzerland) facility[14] and studied for mycotoxin emission due to human excretion. In WWTP effluents, deoxynivalenol was detected in four out of four samples with a mean concentration of 26.1 ng/L (**Table 2.2**). These results are comparable with our earlier study,[14] and proved again the applicability of this method. As for the other mycotoxins, only beauvericin was detected in WWTP effluent, but was not quantifiable.

Thirdly, weekly and fortnightly flow-proportional surface water samples from the Canton of Zurich (Office for Waste, Water, Energy, and Air; AWEL) and by the monitoring program of the Swiss government (National Long-Term Surveillance of Swiss Rivers; NADUF)[25] were analyzed for mycotoxins. Deoxynivalenol was detected in Swiss river water samples with a mean concentration of 5.5 ng/L, which again is comparable with earlier data.[10] Similar to drainage water, beauvericin and nivalenol were also detected in Swiss river waters with mean concentrations of 4.3 ng/L, and 5.9 ng/L, respectively.

Whereas the origin of deoxynivalenol in river water and the relative importance of its sources was recently elucidated by,[14] the presence of other mycotoxins such as nivalenol and beauvericin needs yet to be investigated in further detail. Both nivalenol and beauvericin are primarily field-produced mycotoxins with a considerable aqueous solubility (log Kow deoxynivalenol = - 2.2; EPISuite 4.0; pKa (beauvericin) = - 1.2,[37]). This makes runoff from contaminated fields a likely source, which is supported by the preliminary data presented here. Other compounds like 3-acetyl-deoxynivalenol and zearalenone were detected, but not quantifiable in all of the three different measured environmental matrices. Overall, the necessity to study mycotoxin occurrence, fate and behavior in natural waters is indicated. Whether or not the presence of such compounds in the ng/L up to a few µg/L range poses an ecotoxicological risk in the aqueous environment remains to be investigated with corresponding effect studies.

Table 2.2. Concentrations (nanograms per liter) of the mycotoxins found in drainage water, surface water and WWTP effluent monitored from January to June 2010.

	drainage water				river water				WWTP effluent			
compounds	min-max [ng/L]	mean (STD) [ng/L]	no. of samples	detected in X samples	min-max [ng/L]	mean (STD) [ng/L]	no. of samples	detected in X samples	min-max [ng/L]	mean (STD) [ng/L]	no. of samples	detected in X samples
beauvericin	d-10.5	6.7 (2.0)	129	39	d-52.8	4.3 (1.6)	223	77	d	d	4	4
deoxynivalenol	d-22.5	8.3 (1.9)	129	22	d-11.9	5.5 (2.6)	223	37	16.4-38.8	26.1 (3.2)	4	4
nivalenol	d-8.5	6.1 (3.3)	129	26	d-9.2	5.9 (2.7)	73	35	nd	nd	4	0
3-acetyl-deoxynivalenol	d	d	129	10	d	d	223	93	d	d	4	4
zearalenone	d	d	129	3	d	d	223	7	nd	nd	4	0

d: detected but not quantifiable; nd: not detected; STD: standard deviation

Acknowledgment

We thank the Federal Office for the Environment for the financial support of this project. We gratefully acknowledge P. Niederhauser (AWEL) and B. Luder (NADUF) for providing surface water samples from different sampling stations throughout Switzerland. Additionally, we like to thank the WWTP personnel of Kloten/Opfikon for their help and support during the sampling time. Further, we would also thank S. Vogelgsang and H.R. Forrer for their everlasting enthusiasm to share with us their knowledge on *Fusarium* diseases. Last but not least, we are thankful to the field staff at ART for their great support at the field site.

References (chapter 2)

1. Lanier, C., et al., Mycoflora and mycotoxin production in oilseed cakes during farm storage. *J. Agric. Food Chem.* **2009**, *57*, 1640-1645.
2. Faberi, A., et al., Determination of type B fumonisin mycotoxins in maize and maize-based products by liquid chromatography/tandem mass spectrometry using a QqQ(linear ion trap) mass spectrometer. *Rapid Commun. Mass Spectrom.* **2005**, *19*, 275-282.
3. Hussein, H. S.; Brasel, J. M., Toxicity, metabolism, and impact of mycotoxins on humans and animals. *Toxicology* **2001**, *167*, 101-134.
4. Richard, J. L., Some major mycotoxins and their mycotoxicoses - An overview. *Int. J. Food Microbiol.* **2007**, *119*, 3-10.
5. Kuiper-Goodman, T., Food safety: Mycotoxins and phycotoxins in perspective. In *Mycotoxins and Phycotoxins - Developments in Chemistry, Toxicology and Food Safety*, Miraglia, M.; VanEgmond, H. P.; Brera, C.; Gilbert, J., Eds. Alaken, Inc: Ft Collins, **1998**; pp 25-48.
6. Zollner, P.; Mayer-Helm, B., Trace mycotoxin analysis in complex biological and food matrices by liquid chromatography-atmospheric pressure ionisation mass spectrometry. *J. Chromatogr. A* **2006,** *1136*, 123-169.
7. Krska, R., et al., Analysis of *Fusarium* toxins in feed. *Ani. Feed Sci. Technol.* **2007**, *137*, 241-264.

8. Hartmann, N., et al., Occurrence of zearalenone on *Fusarium graminearum* infected wheat and maize fields in crop organs, soil and drainage water. *Environ. Sci. Technol.* **2008**, *42*, 5455-5460.
9. Hartmann, N., et al., Quantification of zearalenone in various solid agroenvironmental samples using D-6-zearalenone as the internal standard. *J. Agric. Food Chem.* **2008**, *56*, 2926-2932.
10. Bucheli, T. D., et al., *Fusarium* mycotoxins: Overlooked aquatic micropollutants? *J. Agric. Food Chem.* **2008**, *56*, 1029-1034.
11. Leu, C., et al., Simultaneous assessment of sources, processes, and factors influencing herbicide losses to surface waters in a small agricultural catchment. *Environ. Sci. Technol.* **2004**, *38*, 3827-3834.
12. Gilbert, J., et al. In *Assessment of dietary exposure to ochratoxin A in the UK using a duplicate diet approach and analysis of urine and plasma samples*, 1st Symposium on Risk Assessment and Communication for Food Safety, York, England, Jun, 2000; Taylor & Francis Ltd: York, England, **2000**; pp 1088-1093.
13. Turner, P. C., et al., Urinary deoxynivalenol is correlated with cereal intake in individuals from the United Kingdom. *Environ. Health Perspec.* **2008**, *116*, 21 - 25.
14. Wettstein, F. E.; Bucheli, T. D., Poor elimination rates in waste water treatment plants lead to continuous emission of deoxynivalenol into the aquatic environment. *Wat. Res.* **2010**, *44*, 4137-4142.
15. Placinta, C. M., et al., A review of worldwide contamination of cereal grains and animal feed with *Fusarium* mycotoxins. *Animal Feed Science and Technology* **1999**, *78*, 21-37.
16. Batt, A. L.; Aga, D. S., Simultaneous analysis of multiple classes of antibiotics by ion trap LC/MS/MS for assessing surface water and groundwater contamination. *Anal. Chem.* **2005**, *77*, 2940-2947.
17. Wick, A., et al., Comparison of electrospray ionization and atmospheric pressure chemical ionization for multi-residue analysis of biocides, UV-filters and benzothiazoles in aqueous matrices and activated sludge by liquid chromatography-tandem mass spectrometry. *J. Chromatogr. A* **2010**, *1217*, 2088-2103.
18. Cavaliere, C., et al., Mycotoxins produced by *Fusarium* genus in maize: determination by screening and confirmatory methods based on liquid

chromatography tandem mass spectrometry. *Food Chemistry* **2007**, *105*, 700-710.

19. van Bennekom, E. O., et al., Confirmatory analysis method for zeranol, its metabolites and related mycotoxins in urine by liquid chromatography-negative ion electrospray tandem mass spectrometry. *Anal. Chim. Acta* **2002**, *473*, 151-160.
20. Vatinno, R., et al., Automated high-throughput method using solid-phase microextraction-liquid chromatography-tandem mass spectrometry for the determination of ochratoxin A in human urine. *J. Chromatogr. A* **2008**, *1201*, 215-221.
21. Freitas, L. G., et al., Quantification of the new triketone herbicides, sulcotrione and mesotrione, and other important herbicides and metabolites, at the ng/l level in surface waters using liquid chromatography- tandem mass spectrometry. *J. Chromatogr. A* **2004**, *1028*, 277-286.
22. Haubl, G., et al., Characterization of (C-13(24)) T-2 toxin and its use as an internal standard for the quantification of T-2 toxin in cereals with HPLC-MS/MS. *Anal. Bioanal. Chem.* **2007**, *389*, 931-940.
23. Asam, S.; Rychlik, M., Synthesis of four carbon-13-labeled type A trichothecene mycotoxins and their application as internal standards in stable isotope dilution assays. *J. Agric. Food Chem.* **2006**, *54*, 6535-6546.
24. Hartmann, N., et al., Quantification of estrogenic mycotoxins at the ng/L level in aqueous environmental samples using deuterated internal standards. *J. Chromatogr. A* **2007**, *1138*, 132-140.
25. Hoerger, C. C., et al., Occurrence and origin of estrogenic isoflavones in Swiss river waters. *Environ. Sci. Technol.* **2009**, *43*, 6151-6157.
26. Keith, L. H., et al., Principles of environmental-analysis. *Anal. Chem.* **1983**, *55*, 2210-2218.
27. Reemtsma, T.; Quintana, J. B., Analytical Method for Polar Pollutants In *Organic Pollutants in the Water Cycle*, Reemtsma, T.; Jekel, M., Eds. Wiley-VCH Verlag GmbH & Co. KGaA: Weinheim, **2006**; Vol. 1, pp 1-34.
28. Ollers, S., et al., Simultaneous quantification of neutral and acidic pharmaceuticals and pesticides at the low-ng/l level in surface and waste water. *J. Chromatogr. A* **2001**, *911*, 225-234.

29. Vogelgesang, J.; Hädrich, J., Limits of detection, identification and determination: a statistical approach for practitioners. *Acc. Quality Assurrence* **1998**, 242-255.
30. Romero-Gonzalez, R., et al., Application of Conventional Solid-Phase Extraction for Multimycotoxin Analysis in Beers by Ultrahigh-Performance Liquid Chromatography-Tandem Mass Spectrometry. *J. Agric. Food Chem.* **2009**, *57*, 9385-9392.
31. Sulyok, M., et al., Development and validation of a liquid chromatography/tandem mass spectrometric method for the determination of 39 mycotoxins in wheat and maize. *Rapid Commun. Mass Spectrom.* **2006**, *20*, 2649-2659.
32. Commission Decision 2002/657/EC implementing Council Directive 96/23/EC concerning the performance of analytical methods and the interpretation of results Brussels, **2002**.
33. Antignac, J. P., et al., The ion suppression phenomenon in liquid chromatography-mass spectrometry and its consequences in the field of residue. *Anal. Chim. Acta* **2005**, *529*, 129-136.
34. Dijkman, E., et al., Study of matrix effects on the direct trace analysis of acidic pesticides in water using various liquid chromatographic modes coupled to tandem mass spectrometric detection. *J. Chromatogr. A* **2001**, *926*, 113-125.
35. Lagana, A., et al., Development of an analytical system for the simultaneous determination of anabolic macrocyclic lactones in aquatic environmental Samples. *Rapid Commun. Mass Spectrom.* **2001**, *15*, 304-310.
36. Kasprzyk-Hordern, B., et al., The effect of signal suppression and mobile phase composition on the simultaneous analysis of multiple classes of acidic/neutral pharmaceuticals and personal care products in surface water by solid-phase extraction and ultra performance liquid chromatography-negative electrospray tandem mass spectrometry. *Talanta* **2008**, *74*, 1299-1312.
37. Jestoi, M., Emerging Fusarium-mycotoxins fusaproliferin, beauvericin, enniatins, and moniliformin - A review. *Crit. Rev. Food Sci.* **2008**, *48*, 21-49.

Supporting Information

Figure S2.1. Schematic overview of the method validation process A) for all mycotoxins without ILIS and B) with ILIS. For details on the process flow, see text section method validation parameters.

A) Methodvalidationprocess without ILIS

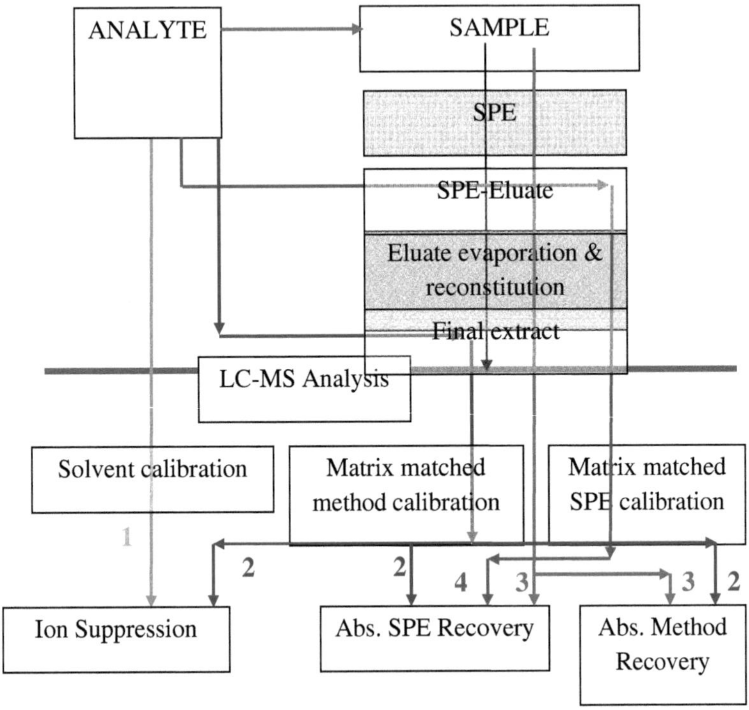

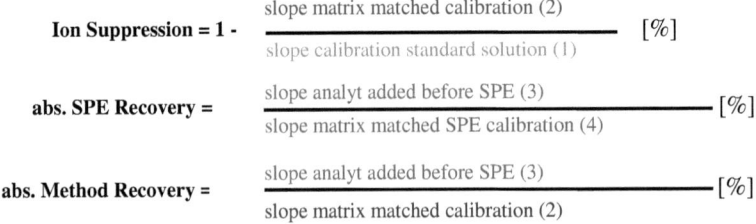

$$\textbf{Ion Suppression} = 1 - \frac{\text{slope matrix matched calibration (2)}}{\text{slope calibration standard solution (1)}} \ [\%]$$

$$\textbf{abs. SPE Recovery} = \frac{\text{slope analyt added before SPE (3)}}{\text{slope matrix matched SPE calibration (4)}} \ [\%]$$

$$\textbf{abs. Method Recovery} = \frac{\text{slope analyt added before SPE (3)}}{\text{slope matrix matched calibration (2)}} \ [\%]$$

B) Methodvalidationprocess with ILIS

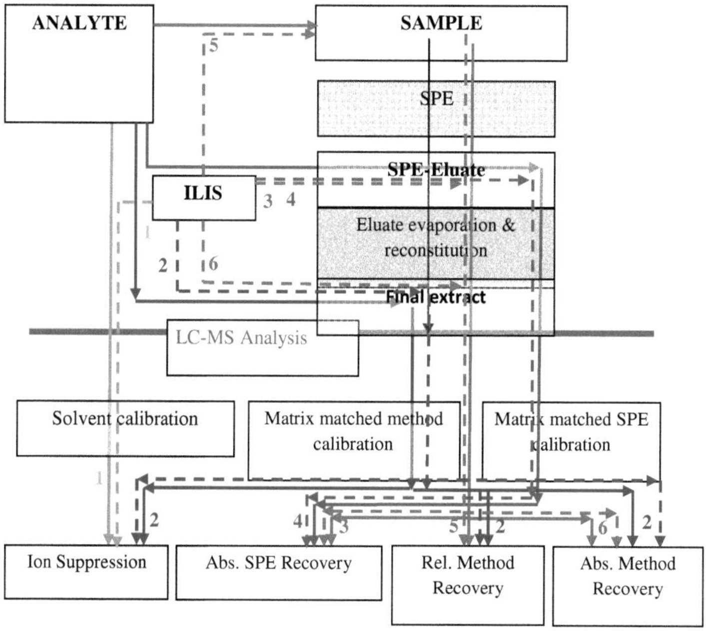

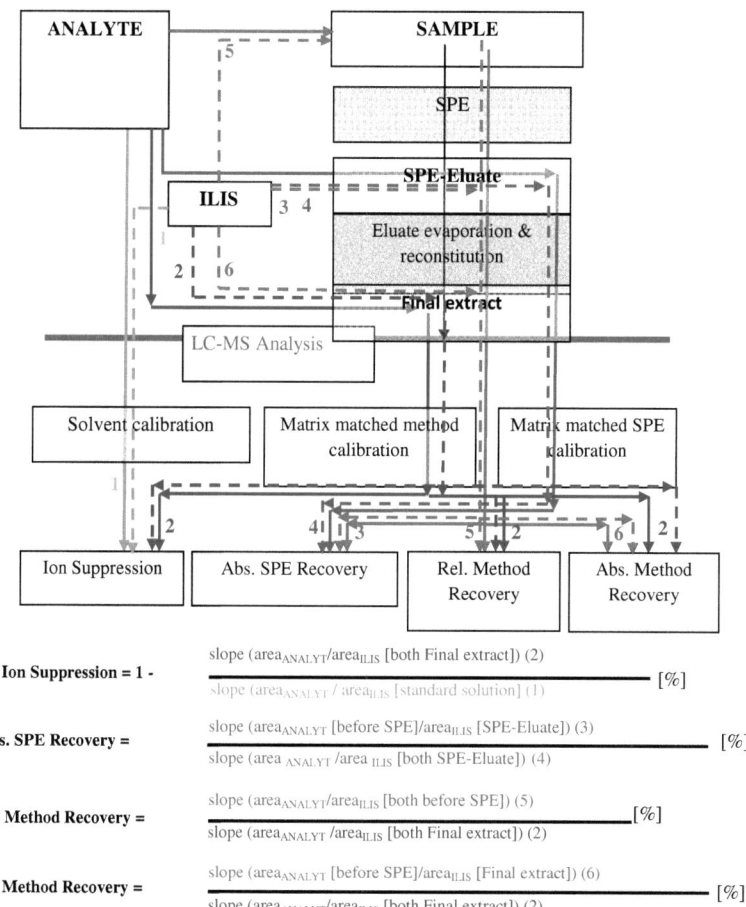

$$\textbf{Ion Suppression} = 1 - \frac{\text{slope (area}_{ANALYT}\text{/area}_{ILIS}\text{ [both Final extract]) (2)}}{\text{slope (area}_{ANALYT}\text{ / area}_{ILIS}\text{ [standard solution] (1)}} \text{ [\%]}$$

$$\textbf{Abs. SPE Recovery} = \frac{\text{slope (area}_{ANALYT}\text{ [before SPE]/area}_{ILIS}\text{ [SPE-Eluate]) (3)}}{\text{slope (area }_{ANALYT}\text{ /area }_{ILIS}\text{ [both SPE-Eluate]) (4)}} \text{ [\%]}$$

$$\textbf{Rel. Method Recovery} = \frac{\text{slope (area}_{ANALYT}\text{/area}_{ILIS}\text{ [both before SPE]) (5)}}{\text{slope (area}_{ANALYT}\text{ /area}_{ILIS}\text{ [both Final extract]) (2)}} \text{ [\%]}$$

$$\textbf{Abs. Method Recovery} = \frac{\text{slope (area}_{ANALYT}\text{ [before SPE]/area}_{ILIS}\text{ [Final extract]) (6)}}{\text{slope (area}_{ANALYT}\text{/area}_{ILIS}\text{ [both Final extract]) (2)}} \text{ [\%]}$$

Figure S2.2. LC-MS/MS chromatograms from WWTP effluents extracts spiked with 25 ng/L of all mycotoxins measured A) in the negative ionisation mode (except beauvericin, but positioned here due to spare place arrangements) and B) in the positive ionisation mode. For details on the instrumental parameters, see text and Table 1 in the manuscript. Name of each mycotoxin and their corresponding quantifier transition are written next to the chromatogram.

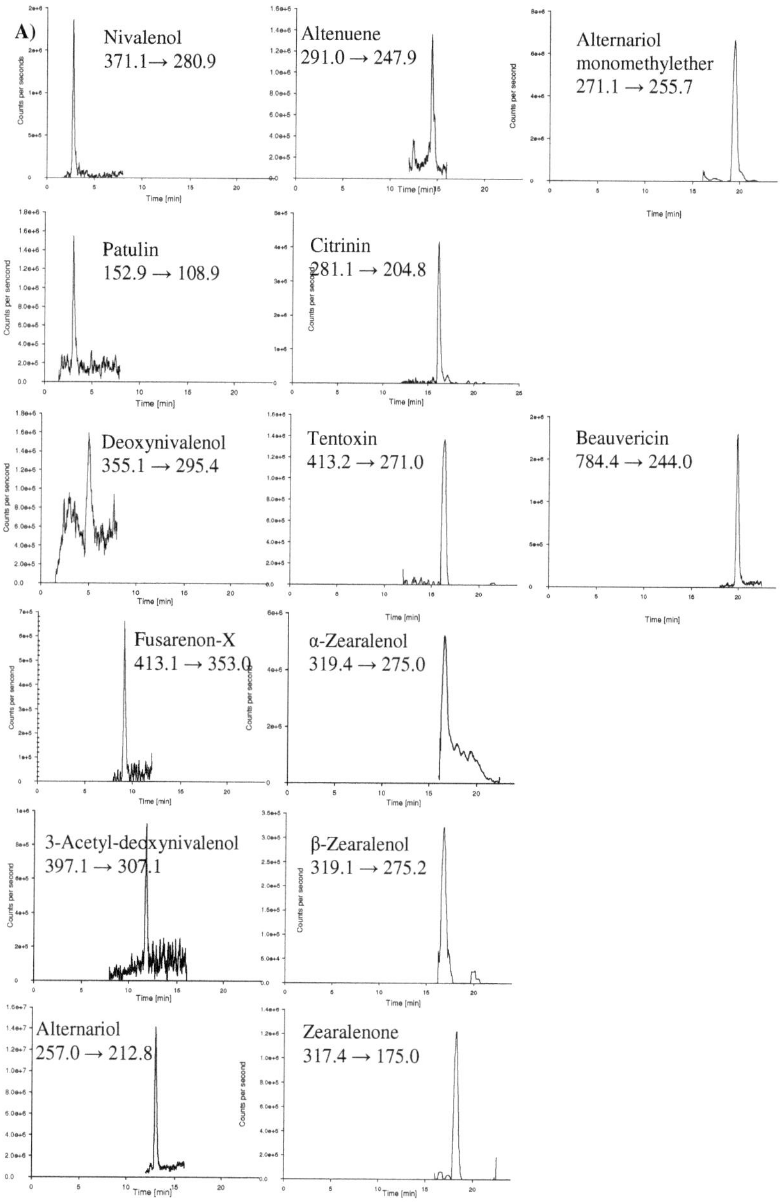
A)
Nivalenol
371.1→ 280.9
Altenuene
291.0 → 247.9
Alternariol
monomethylether
271.1 → 255.7
Patulin
152.9 → 108.9
Citrinin
281.1 → 204.8
Deoxynivalenol
355.1 → 295.4
Tentoxin
413.2 → 271.0
Beauvericin
784.4 → 244.0
Fusarenon-X
413.1 → 353.0
α-Zearalenol
319.4 → 275.0
3-Acetyl-deoxynivalenol
397.1 → 307.1
β-Zearalenol
319.1 → 275.2
Alternariol
257.0 → 212.8
Zearalenone
317.4 → 175.0
Counts per second
Time [min]

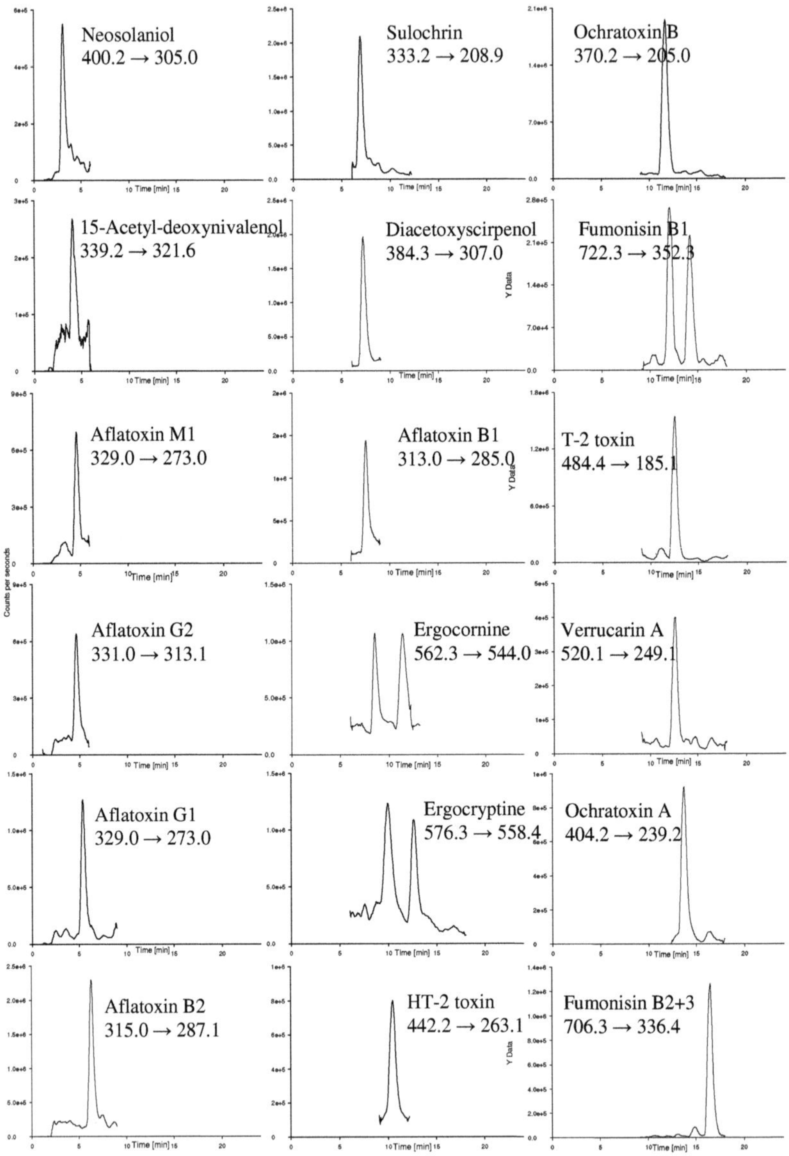
Neosolaniol
400.2 → 305.0
Sulochrin
333.2 → 208.9
Ochratoxin B
370.2 → 205.0
15-Acetyl-deoxynivalenol
339.2 → 321.6
Diacetoxyscirpenol
384.3 → 307.0
Fumonisin B1
722.3 → 352.3
Aflatoxin M1
329.0 → 273.0
Aflatoxin B1
313.0 → 285.0
T-2 toxin
484.4 → 185.1
Aflatoxin G2
331.0 → 313.1
Ergocornine
562.3 → 544.0
Verrucarin A
520.1 → 249.1
Aflatoxin G1
329.0 → 273.0
Ergocryptine
576.3 → 558.4
Ochratoxin A
404.2 → 239.2
Aflatoxin B2
315.0 → 287.1
HT-2 toxin
442.2 → 263.1
Fumonisin B2+3
706.3 → 336.4
Time [min]
Counts per seconds
Y Data

Figure S2.3. Absolute cartridge recovery of A): mycotoxins measured in negative ionisation mode (*-ESI*) and B): mycotoxins measured in positive ionisation mode (*+ESI*) extracted from MilliQ-water. The red line indicates the lower end of the satisfactory recovery range (70%). Abbreviations: BEP: Bond Elut Plexa (VarianInc.); Chroma: Chromabond HR-X (Macherey & Nagel), Oasis HLB (Waters Corporation).

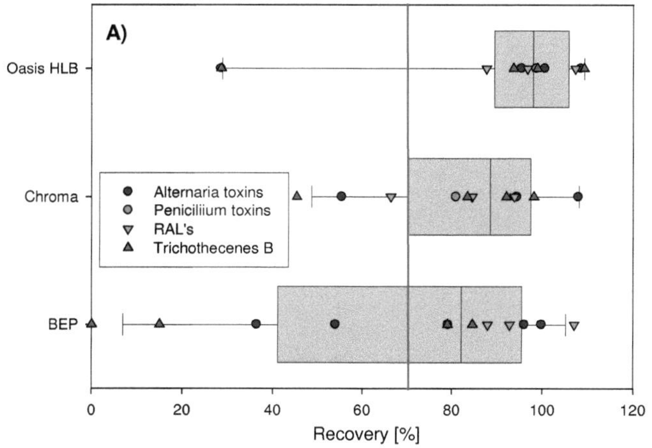

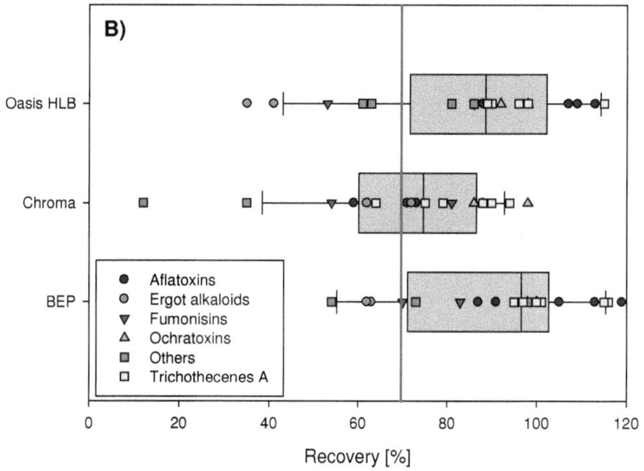

Table S2.1. Absolute SPE recoveries of mycotoxins at 100 ng/L in Milli-Q water.

Compounds	BEP [%]	HR-X [%]	Oasis HLB [%]
Aflatoxins			
AFB_1	87	59	113
AFB_2	91	59	107
AFG_1	113	71	109
AFG_2	119	73	89
AFM_1	105	88	88
Alternaria toxins			
ALT	96	94	95
AME	36	55	29
AOH	54	108	109
Tet	100	108	101
Ergot alkaloids			
ECO	63	62	41
ECY	62	72	35
Fumonisins			
FB_1	70	81	53
FB_{2+3}	83	54	86
Ochratoxins			
OTA	98	96	98
OTB	100	98	92
Others			
BEA	54	35	63
STC	73	12	61
SUL	98	79	81
Penicillium			
CIT	79	81	99
PAT	-	-	29
RAL			
α-ZOL	93	85	88
β-ZOL	87	94	97
ZON	88	66	107
Trichothecene A			
DAS	97	88	98
HT-2	116	90	115
NEO	115	79	90
T-2	100	64	96
VERA	101	75	89
Trichothecene B			
3-Ac-DON	85	98	94
15-Ac-DON	95	94	122
DON	15	92	99
FUS-X	79	83	109
NIV	3	45	29
min. value	3	12	29
max. value	119	108	122
median	**90**	**80**	**94**

n.m. not measurable for this compound

Table S2.2. Ion suppression of mycotoxins in various natural water extracts.

Compounds	Milli-Q-water [%]	Drainage water [%]	River water [%]	WWTP effluent [%]
Aflatoxins				
AFB_1	0	78	40	57
AFB_2	8	68	46	22
AFG_1	14	82	80	73
AFG_2	15	n.a.	73	75
AFM_1	7	76	48	67
Alternaria toxins				
ALT	-5	71	80	87
AME	0	29	31	79
AOH	8	n.a.	n.a.	75
Tet	-27	86	89	94
Ergot alkaloids				
ECO	-5	13	29	66
ECY	-3	9	15	60
Fumonisins				
FB_1	45	52	51	51
FB_{2+3}	9	5	4	-48
Ochratoxins				
OTA	6	7	2	-7
OTB	7	25	26	47
Others				
BEA	-31	15	-19	18
STC	-63	-4	17	42
SUL	10	71	43	95
Penicillium				
CIT	-12	9	17	51
PAT	2	35	96	52
RAL				
α-ZOL	-7	80	84	88
β-ZOL	-12	74	86	84
ZON	21	12	22	57
Trichothecene A				
DAS	3	33	30	53
HT-2	-8	-10	-11	-16
NEO	11	74	60	80
T-2	6	5	-2	-4
VERA	8	37	32	30
Trichothecene B				
3-Ac-DON	-3	-4	-10	-42
15-Ac-DON	18	n.a.	66	n.a.
DON	-14	-14	-8	-61
FUS-X	6	78	79	82
NIV	3	20	39	-28
min. value	0	-4	-2	-4
max. value	-63	99	96	94
median	**3**	**31**	**36**	**55**

n.a. not available due to higher method detection limit

Table S2.3. Absolute method recoveries, method precision (MP) and instrument precision (IP) of mycotoxins at low concentrations in various nat

Compounds	Concentration	Milli-Q-water [%]	MP [%]	IP [%]	Drainage water [%]	MP [%]	IP [%]	River water [%]	MP [%]	IP [%]	WWTP [%]	MP [%]	IP [%]
Aflatoxins													
AFB_1	5	92 (2)	2		67 (6)	10		43 (2)	4		54 (3)	6	
	25	72 (4)	5		52 (0)	1		37 (5)	13		61 (3)	5	
	100	90 (3)	3	3	69 (5)	7	6	51 (7)	13	1	52 (4)	7	2
AFB_2	5	83 (5)	6		41 (8)	18		46 (4)	8		59 (3)	5	
	25	72 (3)	4		52 (4)	7		42 (3)	8		56 (3)	5	
	100	91 (3)	3	1	57 (4)	7	3	55 (5)	9	3	53 (4)	8	5
AFG_1	5	103 (4)	4		n.a.	n.a.		n.a.	n.a.		35 (4)	12	
	25	81 (1)	2		6 (0)	19		n.a.	n.a.		31 (8)	26	
	100	88 (5)	6	2	6 (0)	5	11	6 (1)	16	18	23 (7)	32	5
AFG_2	5	92 (10)	11		n.a.	n.a.		n.a.	n.a.		n.a.	n.a.	
	25	80 (4)	5		n.a.	n.a.		n.a.	n.a.		47 (6)	13	
	100	85 (4)	5	3	n.a.	n.a.	n.m.	9 (2)	27	14	30 (7)	23	10
AFM_1	5	85 (4)	4		n.a.	n.a.		47 (4)	9		46 (6)	14	
	25	79 (2)	2		37 (1)	3		30 (3)	11		57 (6)	11	
	100	86 (3)	3	3	44 (1)	3	8	38 (5)	12	6	48 (2)	5	9
Alternaria toxins													
ALT	5	73 (5)	7		n.a.	n.a.		n.a.	n.a.		n.a.	n.a.	
	25	85 (4)	5		n.a.	n.a.		n.a.	n.a.		n.a.	n.a.	
	100	90 (3)	3	8	89 (1)	1	11	81 (9)	11	2	96 (2)	2	6
AME	5	43 (3)	8		39 (6)	15		14 (2)	11		42 (2)	5	
	25	35 (2)	7		26 (2)	6		24 (3)	12		25 (2)	6	
	100	40 (2)	5	2	23 (0)	1	7	10 (3)	18	5	18 (1)	8	11
AOH	5	90 (6)	7		n.a.	n.a.		n.a.	n.a.		33 (6)	17	
	25	64 (2)	3		n.a.	n.a.		n.a.	n.a.		26 (2)	8	
	100	67 (2)	4	11	n.a.	n.a.	n.m.	n.a.	n.a.	n.m.	21 (1)	11	4
Tet	5	92 (5)	5		64 (7)	19		n.a.	n.a.		n.a.	n.a.	
	25	74 (5)	7		63 (12)	11		50 (8)	16		79 (4)	5	
	100	67 (3)	4	3	64 (2)	4	4	51 (3)	6	8	61 (3)	6	4
Ergot alkaloids													
ECO	5	35 (4)	14		n.a.	n.a.		n.a.	n.a.		n.a.	n.a.	
	25	34 (5)	6		57 (6)	11		58 (4)	7		n.a.	n.a.	
	100	39 (4)	9	4	44 (4)	10	12	42 (3)	6	16	20 (1)	6	9
ECY	5	29 (3)	12		69 (2)	10		62 (6)	3		44 (4)	8	
	25	28 (3)	11		52 (3)	6		52 (3)	6		39 (5)	14	
	100	33 (3)	9	1	41 (4)	10	5	35 (2)	7	9	25 (5)	20	10

Compounds	Concentration	Milli-Q-water [%]	MP [%]	IP [%]	Drainage water [%]	MP [%]	IP [%]	River water [%]	MP [%]	IP [%]	WWTP [%]	MP [%]	IP [%]
Fumonisins													
FB_1	5	n.a.	n.a.		n.a.	n.a.		n.a.	n.a.		n.a.	n.a.	
	25	67 (4)	6		n.a.	n.a.		n.a.	n.a.		29 (8)	27	
	100	48 (4)	9	18	n.a.	n.a.	n.m.	14 (1)	8	20	22 (3)	14	50
FB_{2+3}	5	n.a.	n.a.		n.a.	n.a.		n.a.	n.a.		n.a.	n.a.	
	25	67 (3)	4		n.a.	n.a.		n.a.	n.a.		n.a.	n.a.	
	100	75 (11)	15	10	6 (0)	5	18	22 (4)	20	21	15 (2)	12	12
Ochratoxins													
OTA	5	99 (6)	6		78 (6)	7		78 (5)	6		89 (5)	5	
	25	92 (4)	4		84 (3)	4		93 (5)	5		89 (4)	5	
	100	103 (3)	3	5	87 (2)	2	4	91 (3)	4	4	77 (3)	3	6
OTB	5	93 (5)	5		100 (5)	5		62 (4)	6		72 (4)	6	
	25	87 (4)	4		93 (4)	5		75 (7)	9		76 (2)	3	
	100	94 (4)	4	0	90 (4)	5	1	77 (6)	7	2	75 (2)	3	3
Others													
BEA	5	65 (2)	2		66 (9)	14		56 (8)	14		76 (6)	5	
	25	44 (2)	6		60 (2)	3		73 (11)	15		77 (4)	7	
	100	38 (3)	7	3	84 (10)	12	2	44 (8)	17	1	104 (8)	7	4
STC	5	67 (6)	9		54 (8)	15		37 (4)	10		50 (3)	7	
	25	40 (3)	7		52 (3)	7		47 (6)	12		46 (5)	10	
	100	54 (3)	5	4	69 (3)	5	3	68 (6)	9	3	64 (6)	10	5
SUL	5	81 (11)	13		n.a.	n.a.		n.a.	n.a.		n.a.	n.a.	
	25	80 (4)	4		80 (4)	6		n.a.	n.a.		n.a.	n.a.	
	100	75 (4)	5	4	109 (13)	12	3	35 (5)	15	6	43 (13)	33	8
Penicillium													
CIT	5	86 (2)	3		53 (2)	4		46 (2)	3		74 (2)	3	
	25	86 (2)	2		62 (3)	5		57 (2)	3		72 (2)	3	
	100	88 (3)	4	3	61 (1)	1	2	60 (2)	4	1	59 (1)	2	3
PAT	5	27 (1)	5		n.a.	n.a.		n.a.	n.a.		n.a.	n.a.	
	25	20 (1)	6		15 (4)	24		n.a.	n.a.		n.a.	n.a.	
	100	20 (2)	8	9	12 (0)	2	7	5 (2)	32	5	9 (1)	8	15
RAL													
α-ZOL	5	81 (5)	6		n.a.	n.a.		n.a.	n.a.		n.a.	n.a.	
	25	70 (5)	7		n.a.	n.a.		70 (1)	2		n.a.	n.a.	
	100	90 (3)	3	3	56 (13)	24	16	68 (6)	8	21	14 (6)	24	7
β-ZOL	5	86 (6)	7		n.a.	n.a.		n.a.	n.a.		n.a.	n.a.	
	25	68 (3)	5		n.a.	n.a.		95 (6)	6		21 (7)	34	
	100	94 (3)	3	9	54 (7)	12	29	49 (5)	10	26	29 (9)	33	7

Compounds	Concentration	Milli-Q-water [%]	MP [%]	IP [%]	Drainage water [%]	MP [%]	IP [%]	River water [%]	MP [%]	IP [%]	WWTP [%]	MP [%]	IP [%]
ZON	5	99 (5)	5		n.a.	n.a.		n.a.	n.a.		n.a.	n.a.	
	25	87 (5)	6		104 (7)	7		79 (11)	14		80 (15)	20	
	100	109 (5)	5	4	95 (4)	4	7	83 (6)	8	5	74 (9)	6	23
Trichothecene A													
DAS	5	86 (1)	1		79 (2)	3		64 (2)	3		75 (3)	4	
	25	85 (2)	2		100 (1)	1		71 (1)	2		72 (2)	2	
	100	91 (4)	4	1	95 (1)	1	1	71 (2)	3	1	63 (1)	1	3
HT-2	5	73 (4)	5		90 (8)	9		n.a.	n.a.		n.a.	n.a.	
	25	95 (7)	7		87 (2)	3		96 (9)	9		90 (6)	6	
	100	96 (1)	1	5	86 (2)	2	5	89 (6)	7	8	74 (2)	3	6
NEO	5	86 (7)	8		n.a.	n.a.		79 (8)	10		72 (7)	9	
	25	75 (3)	4		31 (1)	3		52 (7)	13		75 (7)	9	
	100	83 (1)	2	2	35 (2)	5	5	57 (4)	7	4	70 (2)	3	6
T-2	5	104 (4)	4		89 (5)	6		80 (7)	9		94 (7)	8	
	25	91 (2)	2		92 (2)	2		91 (5)	5		85 (6)	7	
	100	100 (5)	5	2	95 (3)	3	2	88 (2)	3	3	80 (1)	2	8
VERA	5	89 (14)	16		n.a.	n.a.		n.a.	n.a.		72 (4)	6	
	25	88 (7)	8		70 (6)	9		70 (2)	2		76 (12)	16	
	100	97 (3)	3	5	63 (2)	3	9	68 (3)	4	3	73 (5)	7	5
Trichothecene B													
3-Ac-DON	5	98 (7)	7		n.a.	n.a.		n.a.	n.a.		n.a.	n.a.	
	25	78 (9)	11		99 (7)	7		75 (7)	10		79 (7)	9	
	100	78 (3)	3	6	96 (5)	5	8	77 (6)	8	7	85 (2)	2	17
15-Ac-DON	5	n.a.	n.a.		n.a.	n.a.		n.a.	n.a.		n.a.	n.a.	
	25	102 (6)	6		n.a.	n.a.		n.a.	n.a.		n.a.	n.a.	
	100	88 (5)	6	9	n.a.	n.a.	n.m.	71 (7)	10	16	n.a.	n.a.	n.m.
DON	5	84 (8)	9		n.a.	n.a.		91 (4)	4		33 (8)	24	
	25	86 (2)	2		99 (4)	4		104 (3)	3		69 (7)	10	
	100	96 (4)	4	6	99 (3)	3	5	94 (6)	7	6	72 (3)	4	9
FUS-X	5	65 (4)	6		n.a.	n.a.		n.a.	n.a.		n.a.	n.a.	
	25	72 (3)	4		88 (5)	6		60 (2)	3		75 (1)	1	
	100	77 (6)	7	2	81 (3)	3	4	62 (2)	4	6	74 (5)	7	4
NIV	5	16 (9)	15		59 (10)	17		n.a.	n.a.		n.a.	n.a.	
	25	17 (4)	22		58 (3)	5		22 (2)	8		20 (1)	6	
	100	24 (3)	11	9	55 (4)	7	3	21 (2)	10	7	11 (0)	4	10
min. value (5/25/100)		16/17/20	1/2/1	1	39/15/6	3/1/1	1	14/24/5	3/2/3	1	33/20/11	3/1/1	2
max. value (5/25/100)		103/102/103	16/22/15	18	00/104/10	17/24/12	29	91/104/94	14/15/32	26	90/94/104	24/34/33	50
median (5/25/100)		77/70/76	5	4	67/62/62	5	5	59/65/56	8	6	59/71/57	7	7

n.a. not available due to higher method detection limit; [a] Absolute standard deviation (five replicates) in parenthesis; MP: method precision; IP: instrument precisic

Table S2.4. Relative method recoveries of mycotoxins at low concentrations in various natural waters.

Compounds	Concentration [ng/L]	Milli-Q-water [%]	RSD [%]	Drainage water [%]	RSD [%]	River water [%]	RSD [%]	WWTP [%]	RSD [%]
3-Ac-DON	5	106 (8)	7	n.a.	n.a.	n.a.	n.a.	n.a.	n.a.
	25	97 (1)	1	109 (3)	3	76 (7)	9	104 (7)	6
	100	82 (3)	4	101 (2)	2	87 (7)	8	94 (5)	6
DON	5	96 (7)	7	n.a.	n.a.	100 (6)	6	106 (12)	12
	25	96 (1)	1	99 (4)	4	109 (2)	1	105 (5)	4
	100	101 (2)	2	106 (2)	2	106 (3)	3	96 (6)	7
FB_1	5	n.a.	n.a.	n.a.	n.a.	n.a.	n.a.	n.a.	n.a.
	25	62 (14)	23	n.a.	n.a.	n.a.	n.a.	96 (14)	15
	100	87 (9)	10	n.a.	n.a.	76 (26)	35	72 (9)	12
HT-2	5	88 (6)	6	81 (6)	7	n.a.	n.a.	n.a.	n.a.
	25	101 (5)	5	96 (3)	3	78 (8)	11	90 (6)	7
	100	104 (3)	3	102 (6)	5	87 (4)	5	86 (2)	2
OTA	5	105 (3)	3	95 (13)	14	98 (9)	9	99 (4)	4
	25	99 (3)	3	103 (6)	6	101 (3)	3	105 (2)	2
	100	107 (2)	2	104 (2)	2	96 (2)	2	90 (2)	3
T-2	5	124 (6)	5	113 (6)	6	100 (6)	6	99 (13)	13
	25	107 (3)	3	104 (3)	3	102 (5)	5	102 (3)	3
	100	104 (2)	2	109 (2)	2	101 (2)	2	94 (5)	5
ZON	5	134 (5)	4	n.a.	n.a.	n.a.	n.a.	n.a.	n.a.
	25	115 (2)	2	97 (4)	4	101 (6)	6	86 (15)	18
	100	129 (1)	1	102 (1)	1	91 (7)	7	103 (13)	12
min. value (5/25/100)		88/62/82	3/1/1	81/96/101	6/3/1	98/76/76	6/1/2	99/86/72	4/2/2
max. value (5/25/100)		134/115/129	7/23/10	113/109/109	14/6/5	100/109/106	9/11/35	106/105/103	13/18/12
median		**103**	**3**	**102**	**3**	**99**	**6**	**96**	**6**

n.a. not available due to higher method detection limit; [a] Absolute standard deviation (five replicates) in parenthesis

Table S2.5. Instrument detection limit (IDL) and method detection limit (MDL) of mycotoxins at low concentrations in various natural waters.

Compounds	IDL[†] [ng/L]	Milli-Q-water [ng/L]	Drainage water [ng/L]	River water [ng/L]	WWTP effluent [ng/L]
Aflatoxins					
AFB_1	0.1	0.4	4.4[b]	0.3	2.3[b]
AFB_2	0.2	0.8	5.4[b]	0.6	2.1[b]
AFG_1	0.4	0.7	0.6[b]	3.2[c]	6.9[b]
AFG_2	0.9	1.8	n.a.	8.0[c]	5.2[b]
AFM_1	0.4	0.6	10.7[b]	0.7	5.1[b]
Alternaria toxins					
ALT	0.6	0.8	3.8[c]	29[c]	7[c]
AME	0.2	0.5	1.3[b]	2.5[b]	1.4[b]
AOH	0.4	1.1	n.a.	n.a.	1.7[b]
Tet	0.2	0.8	10[b]	6.6[b]	3.6[b]
Ergot alkaloids					
ECO	0.7	0.8	5.1[b]	3.5[b]	3.9[c]
ECY	0.4	0.6	1.7	1.8	2
Fumonisins					
FB_1	1.4	3.6[b]	3.0[c]	3.7[c]	6.5[b]
FB_{2+3}	1.1	2.4[b]	1.0[c]	14.7[c]	6[c]
Ochratoxins					
OTA	0.1	0.9	1	0.8	0.8
OTB	0.1	0.8	0.8	0.6	0.7
Others					
BEA	0.2	0.4	1.4	1.3	3.4
STC	0.3	1	2.9[b]	0.6	4[b]
SUL	1.2	1.8	3.7[b]	18.2[c]	47.7[c]
Penicillium					
CIT	0.1	0.4	0.4	0.3	0.4
PAT	0.2	0.2	6.2[b]	5.6[c]	2.5[c]
RAL					
α-ZOL	0.2	0.8	44.9[c]	18.6[c]	21.5[c]
β-ZOL	0.3	1.1	22.3[c]	4.9[b]	31.1[c]
ZON	0.3	0.9	6[b]	12.3[b]	12.9[b]
Trichothecene A					
DAS	0.1	0.2	0.3	0.4	0.5
HT-2	0.6	0.6	1.6	7.2[b]	4.8[b]
NEO	0.3	1.1	0.8[b]	1.3	5.5[b]
T-2	0.2	0.7	0.8	1.2	1.2
VERA	1.0	2.4	4.7[b]	1.4[b]	10.2[b]
Trichothecene B					
3-Ac-DON	0.9	1.1	5.5[b]	6.1[b]	6[b]
15-Ac-DON	5.1	5.2[b]	n.a.	22.9[c]	n.a.
DON	0.9	1.3	0.8	1.3	1.2
FUS-X	0.6	0.6	4.4[b]	1.3[b]	0.7[b]
NIV	0.1	1.5	1.7	1.5[b]	1.6[c]
min. value	0.1	0.2	0.3	0.3	0.4
max. value	5.1	5.2	44.9	29	47.7
median	**0.3**	**0.8**	**0.9**	**1.0**	**1.0**

[†] IDL calculated from signal to noise ratio of solvent standards; n.a. not available due to higher method detection limit; [a] Three times the absolute standard deviation at 5 ng/L; [b] Three times the absolute standard deviation at 25 ng/L; [c] Three times the absolute standard deviation at 100 ng/L.

CHAPTER 3:

DEVELOPMENT, VALIDATION AND APPLICATION OF A MULTI-MYCOTOXIN METHOD FOR THE ANALYSES OF WHOLE WHEAT PLANTS

JUDITH SCHENZEL,[1,2] HANS-RUDOLF FORRER,[1] SUSANNE VOGELGSANG,[1] AND THOMAS D. BUCHELI[1]

[1] AGROSCOPE RECKENHOLZ-TANIKON, RESEARCH STATION ART, CH-8046 ZURICH, SWITZERLAND

[2] INSTITUTE OF BIOGEOCHEMISTRY AND POLLUTANT DYNAMICS, SWISS FEDERAL INSTITUTE OF TECHNOLOGY, CH-8092 ZURICH, SWITZERLAND

REPRODUCED WITH PERMISSION FROM:

MYCOTOXIN RESEARCH, 2012

VOLUME 28, ISSUE 2, PAGES 135–147

Abstract

Mycotoxins are known to affect the health of humans and husbandry animals. In contrast to wheat grains used for food and feed, whole wheat plants are rarely analyzed for mycotoxins, although contaminated straw could additionally expose animals to these toxic compounds. Since the entire wheat plant may also act as source of mycotoxins emitted into the environment, an analytical method was developed, optimized and validated method for the analysis of 28 different mycotoxins in above-ground material from whole wheat plants. The method comprises solid-liquid extraction and a clean-up step using a Varian Bond Elut Mycotoxin® cartridge, followed by liquid chromatography with electrospray ionization and triple quadrupole mass spectrometry. Total method recoveries for 26 out of 28 compounds were between 69 and 122% and showed limits of detection from 1 - 26 $ng/g_{\text{dry weight (dw)}}$. The overall repeatability for all validated compounds was on average 7%, and their mean ion suppression 65%. Those rather high matrix-effects made it necessary to use matrix-matched calibrations to quantify mycotoxins within whole wheat plants. The applicability of this method is illustrated with data from a winter wheat test field to examine the risks of environmental contamination by toxins following artificial inoculation with four different *Fusarium* species each. The selected data originate from samples of a part of the field which was inoculated with *Fusarium crookwellense.* In the wheat samples, various trichothecenes (3-acetyl-deoxynivalenol, deoxynivalenol, diacetoxyscirpenol, fusarenone-X, nivalenol, HT-2 toxin, and T-2 toxin) as well as beauvericin and zearalenone were identified with concentrations ranging from 32 ng/g_{dw} to 12 x 10^3 ng/g_{dw}.

3.1. Introduction

Mycotoxins are naturally occurring metabolites produced by various fungal species, which can occur in different types of agricultural commodities, such as cereals, fruits and vegetables, but also in by-products such as straw or plant debris remaining in the field after harvest as well as in organic soil material.[1] Fungal infection and mycotoxin contamination either take place pre-harvest in the field or post-harvest during storage. Although many hundreds of fungal toxins are known, a more limited number is generally considered to be of concern. Fungal species of the genera *Alternaria*, *Aspergillus*, *Claviceps*, *Fusarium* and *Penicillium* are commonly known to produce a large array range of mycotoxins with a great high structural diversity (see **Figure 3.1** for some examples). Among the most prominent substance classes are alternaria toxins, aflatoxins, ergot alkaloids, fumonisins, ochratoxins, resorcyclic acid lactones (RALs) and trichothecenes.[2,3] These toxic metabolites are commonly found in crops grown and stored for human or animal consumption, as well as in processed food,[4,5] and their occurrence in food and feed has been studied extensively.[6,7] The presence of mycotoxins in food products even in small amounts can lead to severe toxic effects, which range from acute to chronic symptoms.[8] Several of them (e.g. aflatoxins, ochratoxins and fumonisins) have been ranked as the most important chronic dietary risk factor, higher than synthetic contaminants, plant toxins, food additives or pesticide residues.[9] Some of those fungal metabolites were shown to be mutagenic, teratogenic and/or carcinogenic.[10-12]

Fungal infections in the field, for example by *Fusarium* species, do not only lead to a mycotoxin contamination of the grains, but also affect other vegetative parts of the plants, like stems, leaves, nodes and internodes. Therefore, infestation results also in the production of mycotoxins in different plant organs tissue as well as in straw.[13-15] An obvious source for mycotoxin exposure to husbandry animals is represented by grains, whereas straw is often neglected although it acts as may be an additional contamination route.[16] For example, straw is frequently used as litter in ruminant and pig fattening and breeding, and thus might contribute to mycotoxin exposure in these animals.[17] Interestingly, whole wheat plants can additionally act as an environmental input source, as mycotoxin emission by runoff and drainage water from small grain cereal and maize fields was recently confirmed for selected mycotoxins.[18-21]

Generally, the analysis of mycotoxins with one multi-residue method is challenging due to their widely varying chemical and physical properties. However, due to the fact that several mycotoxins co-occur in sometimes trace concentrations within complex matrices, sensitive, reliable and reproducible methods for their quantification are required. Important steps to achieve low detection limits are the sample preparation, extraction and sample clean-up, which are often very time consuming. Different techniques to extract and preconcentrate the sample including solid-liquid extraction (SLE), pressurized liquid extraction (PLE), and accelerated solvent extraction (ASE) have been used.[22-24] Extract clean-up with solid-phase cartridges has been shown to be a suitable tool to improve selectivity and sensitivity in the analysis.[25-28] As a method of choice for separation and detection of mycotoxins in various biological matrices, high performance liquid chromatography electrospray tandem mass spectrometry (HPLC-ESI-MS/MS) has been frequently applied in the recent years.[7,29] Additionally, recent developments in liquid chromatography (LC) allow for even higher resolution and selectivity, as described in Jin et al.[28] for a simultaneous determination of ten mycotoxins using ultra performance LC (UPLC). Nonetheless, one of the drawbacks of these two widely acknowledged techniques is signal suppression or enhancement (SSE), which is caused by co-extracted matrix components. SSE has a profound influence on accuracy, precision and sensitivity, when quantifying low contents of mycotoxins in any sample by HPLC-ESI-MS/MS. Commonly used methods to compensate for SSE are matrix-matched calibration or standard addition.[14,21,27,30-32]

In this study, the development of an accurate, precise and sensitive analytical method for 28 mycotoxins from five fungal genera, representing their major compound classes, in whole wheat plants is described. The selection of mycotoxins was based on several factors, like frequency of occurrence in food and feed matrices, the chemical and physical properties, as well as the toxicity of the compound. The method comprises SLE, filtration, elution over a clean-up cartridge as well as separation and detection with HPLC-MS/MS (ESI +/-). To our knowledge, this is the first method for the quantification of mycotoxins in above-ground material from whole wheat plants. The method was validated using matrix-matched calibration and its application was demonstrated with a selection of initial data from our currently ongoing field study.

3.2. Materials and Methods

3.2.1 Chemicals

3-acetyl-deoxynivalenol (3-AcDON; purity not available [n.a.]), aflatoxin B_1 (AFB_1; ≥99%), aflatoxin B_2 (AFB_2; ≥98%), aflatoxin G_1 (AFG_1; ≥99%), aflatoxin G_2 (AFG_2; ≥99%), aflatoxin M_1 (AFM_1; ≥99%), citrinin (CIT; ≥99%), deoxynivalenol (DON; ≥99%), HT-2 toxin (HT-2; ≥99%), patulin (PAT; ≥99%), and T-2 toxin (T-2; ≥99%) were supplied by Fermentek (Jerusalem, Israel). Alternariol (AOH; ≥96%), alternariol monomethylether (AME; purity [n.a.]), altenuene (ALT; purity [n.a.]), beauvericin (BEA; ≥97%), diacetoxyscirpenol (DAS; ≥99%), ergocornine (ECO; ≥95%), fusarenone-X (FUS-X; purity [n.a.]), neosolaniol (NEO; purity [n.a.]), nivalenol (NIV; purity [n.a.]), sterigmatocystin (STC; ≥98%), sulochrin (SUL; ≥98%), tentoxin (TET; ≥99%), verrucarin A (VERA; ≥98%), zearalenone (ZEA; ≥99%), α-zearalenol (α-ZOL; ≥98%), and β-zearalenol (β-ZOL; ≥95%) were supplied by Sigma-Aldrich GmbH (Buchs, Switzerland). Ergocryptine (ECY; purity [n.a.]) was purchased from Biopure (Referenzsubstanzen GmbH, Tulln, Austria). For structures, see **Figure 3.1**.

Methanol (MeOH, HPLC-grade) and acetonitrile (MeCN, HPLC-grade) were purchased from Scharlau (Barcelona, Spain). Ammonium acetate (*pro analysi*) and acetic acid (*pro analysi*) were supplied by Fluka AG (Buchs, Switzerland). Deionised water was further cleaned with a Milli-Q gradient A10 water purification system from Millipore (Volketswil, Switzerland). High-purity N_2 (99.99995%) and Ar (99.99999%) were obtained from PanGas (Dagmarsellen, Switzerland).

Mycotoxin crystalline compounds were dissolved in pure MeCN, except for FUS-X, NEO, OTA and OTB, which were dissolved in a MeCN/Milli-Q water (50:50, v/v) mixture. The individual stock solutions contained concentrations ranging from 100 to 1000 μg/mL and were used within six month. Multicomponent stock solutions were prepared in MeOH in concentrations of 5 μg/mL. Stock solutions and dilutions thereof were stored in the dark at -20°C and +4°C, respectively.

Aflatoxins [AFs] including Aflatoxin B_1, Aflatoxin B_2, Aflatoxin M_1, Aflatoxin G_1, Aflatoxin G_2

Aflatoxins	R
Aflatoxin B_1	H
Aflatoxin M_1	OH

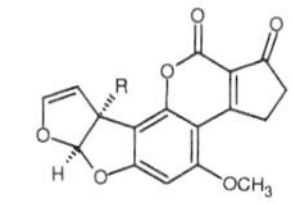

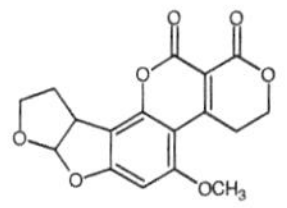

Aflatoxin B_2

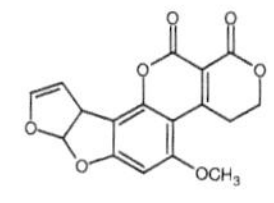

Aflatoxin G_1

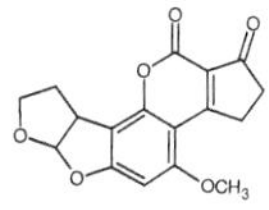

Aflatoxin G_2

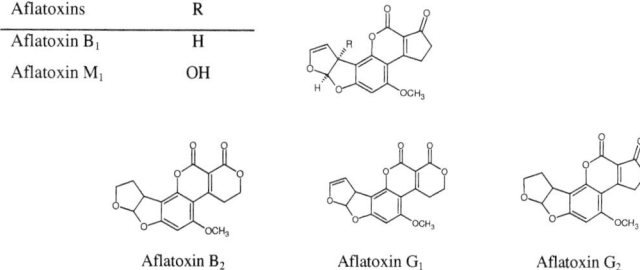

Alternaria toxins:

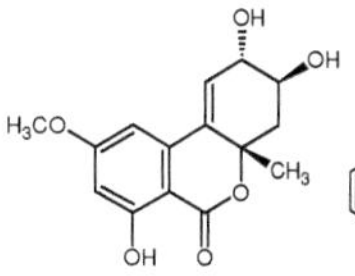

Altenuene

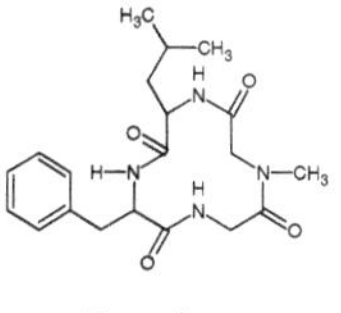

Tentoxin

Alternariol R=OH

Alternariolmethylether R=OCH_3

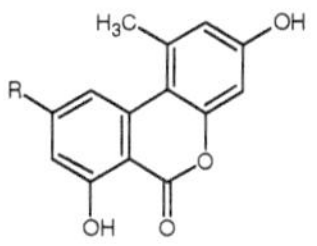

Alternariol

Ergot alkaloids:

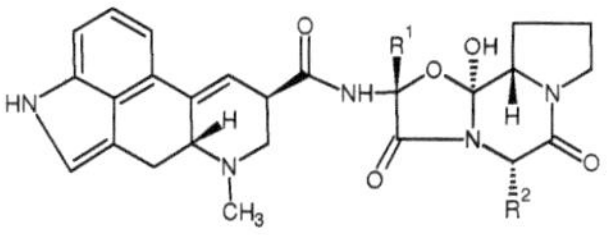

Analyte	R^1	R^2
Ergocornine	$CH(CH_3)_2$	$CH(CH_3)_2$
Ergocryptine	$CH(CH_3)_2$	$CH_2CH(CH_3)_2$

"Other" toxins:

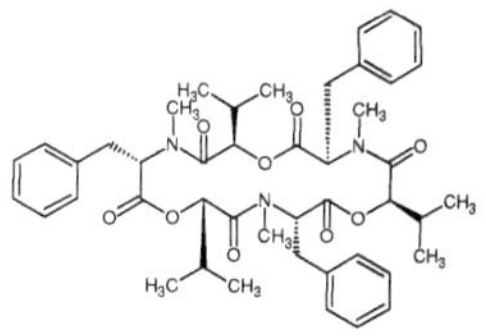

Beauvericin

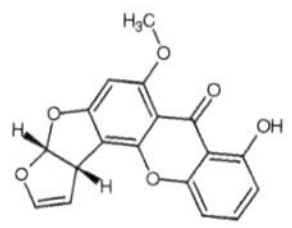

Sterigmatocystin

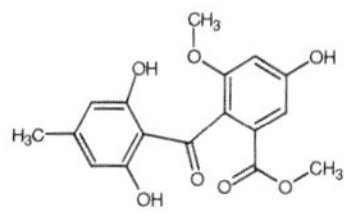

Sulochrin

Penicillium toxins:

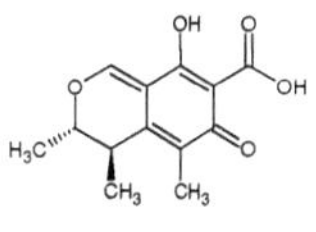

Citrinin

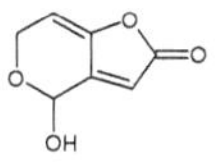

Patulin

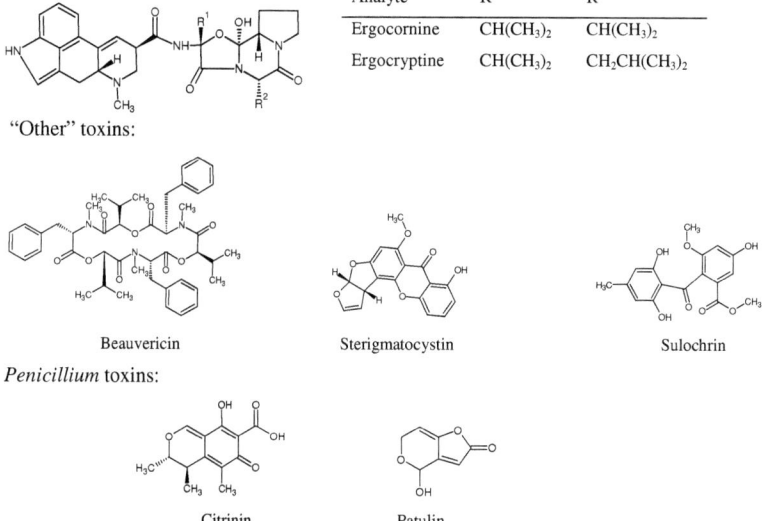

Figure 3.1. Chemical structures of the investigated mycotoxins.

Trichothecenes:

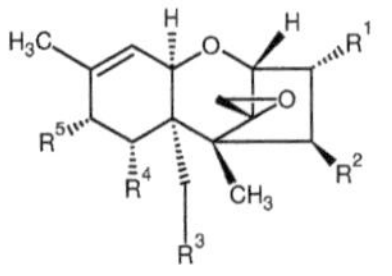

Trichothecene	R^1	R^2	R^3	R^4	R^5
Type A					
DAS	OH	$OCOCH_3$	$OCOCH_3$	H	H
HT-2	OH	OH	$OCOCH_3$	H	$OCOCH_2CH(CH_3)_2$
NEO	OH	$OCOCH_3$	$OCOCH_3$	H	OH
T-2	OH	$OCOCH_3$	$OCOCH_3$	H	$OCOCH_2CH(CH_3)_2$
Type B					
DON	OH	H	OH	OH	=O
FUS-X	OH	$OCOCH_3$	OH	OH	=O
NIV	OH	OH	OH	OH	=O
3-AcDON	$OCOCH_3$	H	OH	OH	=O

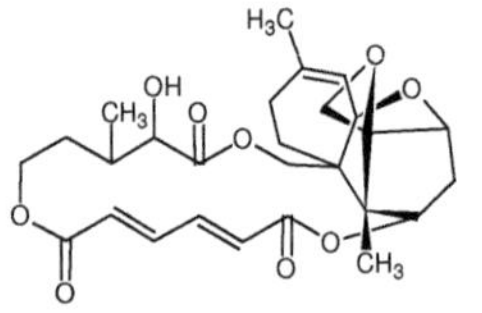

Verrucarin A

Resorcyclic acid lactones:

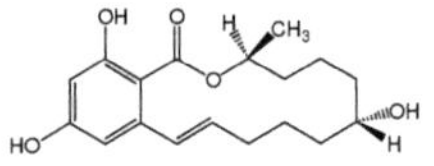

α-Zearalenol

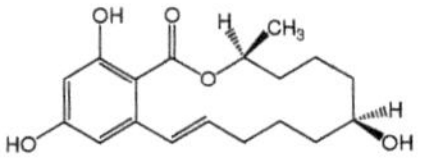

β-Zearalenol

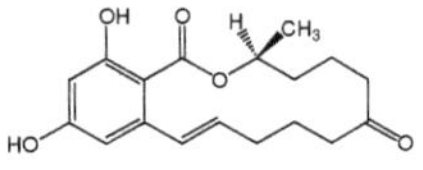

Zearalenone

Figure 3.1. continued.

3.2.2 Field description, sample collection, and sample preparation

The study site (47°,25′,45″ N, 8°,30′,53″ E) was located near the agricultural research station in Reckenholz, north of Zurich, Switzerland. It was divided into two 0.2 ha plots with a gentle slope of 1-2°. The topsoil was classified as a gleyic cambisol (World Reference Base for Soil Resources, FAO, Rome, 1998) with 30% sand (2 mm to 63 mm), 39% silt (<63-2 μm), 31% clay (<2 μm), and a mass fraction of soil organic carbon of 2%. During the whole investigation period air temperature and precipitation data were recorded in intervals of 10 min by the weather station of the Federal Office of Meteorology and Climatology MeteoSwiss (Reckenholz, 443 m above sea level, 47°,25′,40″ N, 8°,31′,04″ E, MeteoSwiss) 300 m near the field site. The study site was managed according to standard agronomic measures except that no fungicides were applied during the flowering of the wheat.

All samples originated from the upper 0.2 hectare field site on which winter wheat of the variety Levis was cultivated from 2008 until 2011. In the year 2008, wheat was artificially contaminated with a *Fusarium avenaceum* conidia suspension (2.5 x 10^5 conidia/mL) at the time of flowering (date depending on weather conditions, 13.06.2008). For method development, whole above-ground wheat plant material was collected manually and randomly within the plot in 2008. In 2009, no samples were collected. In 2010 and 2011, the upper field was divided into several parts, each of which was artificially infected with conidia suspensions of *F. avenaceum*, *F. crookwellense*, *F. graminearum* and *F. poae* at the time of flowering (14.06.2010, and 26.05.2011). For the method application, wheat samples were randomly collected several times before and at the time of harvest in 2010 and 2011 across the parts infected with *F. crookwellense* (exact dates listed in **Table 3.4.**). All samples were first freeze dried at -20°C for several days and then further dried at 35°C until weight consistency was achieved, but at least for 72 h. Afterwards, the samples were ground and sieved to 2 mm using a SM 2000 cutting mill (Retsch GmbH, Haan, Germany). Method application samples were processed within 48h after drying.

3.2.3 Sample extraction, and clean-up

Before extraction, the ground whole-wheat-plant-powder was homogenized with a rotating mixing machine (Turbula® Type T2 F, Willy A. Bachofen AG, Basel, Switzerland) for 15 min. Solid-liquid extraction of 2 g plant material with 20 mL MeCN/Milli-Q water (80:20, v/v) was carried out for 2 h on a SM 30 orbital shaker at 200 rpm (Edmund Bühler GmbH, Hechingen,

Germany). The extraction recoveries of this solvent mixture was compared with those of two other commonly used mixtures, i.e., MeCN/Milli-Q water/acetic acid (79:20:1, v/v/v), and MeCN/Milli-Q water/acetone (50:25:25, v/v/v).[31, 33]

After extraction, the suspended material was filtered over a paper filter No. 595½ with a diameter of 110 mm (Schleicher & Schuell GmbH, Dassel, Germany). For clean-up, all of the filtrate was passed through a Varian Bond Elut Mycotoxin® cartridge (VarianInc. Lake Forest CA, USA) and the eluate was collected in 10 mL conical reaction vial vessels (Supelco, Bellfonte PA, USA). Two other cartridges (R-Biopharm Rhône LTD, Trichothecene EP, Glasgow, Scotland; Romer Labs MycoSep Trichothecene Cleanup, Coring System Diagnostix GmbH, Gernsheim/Rhein, Germany) were also tested for their analyte recovery and purification efficiency. Afterwards, 1 mL of each extract was transferred into 5 mL conical reaction vial vessels and the aliquots were reduced to 100 ± 5 µL under a gentle nitrogen gas stream at 50°C. The extracts were reconstituted in 900 µL of Milli-Q water/MeOH (90:10, v/v) and transferred into 1.5 mL amber glass vials. The samples were stored at +4°C and analyzed within 48 h.

3.2.4 LC-MS/MS Analysis

LC-MS/MS was performed on a Varian 1200L LC-MS instrument (VarianInc, Walnut Creek CA, USA). A detailed description of the LC-MS/MS analysis is presented elsewhere.[21] The mycotoxins were separated using a 125 mm × 2.0 mm i.d., 3 µm, Pyramid C_{18} column, with a 2.1 mm x 20 mm i.d. guard column of the same material (Macherey-Nagel GmbH & Co. KG, Düren, Germany) at room temperature. Two different chromatographic runs were used for the separation of all compounds, one in the positive and one in the negative ionization mode, respectively. The LC mobile phase gradient for the analysis of the analytes measured in negative ionization mode was as follows: 0.0 min: 0 % B (100 % A), 2.0 min: 0 % B, 15.0 min: 100 % B, 18.0 min: 100 % B, 19.0 min: 0 % B, 24.0 min: 0 % B. The analytes measured in the positive ionization mode were separated with the following gradient: 0.0 min: 27% B (73% A), 1.0 min: 27 % B, 1.3 min: 45 % B, 5.0 min: 45 % B, 5.3 min: 63 % B, 9.0 min: 63 % B, 9.3 min: 81 % B, 13.0 min: 81 % B, 13.3 min: 100 % B, 20.0 min: 100 % B, 21.0 min: 27 % B, 24.0 min: 27 % B. In both cases eluent A consisted of Milli-Q water/MeOH/acetic acid (89:10:1, v/v/v) and eluent B of Milli-Q water/MeOH/acetic acid (2:97:1, v/v/v). Both eluents were buffered with 5 mM ammonium acetate. The injection volume was 40 µL and the mobile phase flow was 0.2 mL/min.

LC-MS interface conditions for the ionization of the mycotoxins in the ESI- mode were: needle voltage - 4000 V, nebulizing gas (compressed air) 3.01 bar, drying gas (N_2, 99.5%) 275°C and 1.24 bar, shield voltage -600 V. The neutral mycotoxins were ionized in the ESI+ mode: needle voltage + 4500 V, nebulizing gas (compressed air) 3.01 bar, drying gas (N_2, 99.5%) 275°C and 1.24 bar, shield voltage + 600 V. A detailed list of the analyte-dependent MS/MS parameters and collision cell energies is given in **Table 3.1**. Data processing was carried out using the Varian MS Workstation software (v. 6.9.2.). Quantification took place with matrix-matched calibrations, produced freshly for each batch of samples. Matrix-matched calibration curves were obtained by carrying out standard addition of the multicomponent stock solution to the final extracts to achieve concentrations equivalent to 0, 50, 100, and 150 ng/g. Curves were obtained by plotting measured analyte peak areas against corresponding analyte concentrations in the extracted matrix. Linear regression was performed for each curve. Positive identification criteria for the target analytes were their relative retention time, and their main-to-second product ion ratio (**Table 3.1**).

Table 3.1. Optimized analytical parameters for the selected mycotoxins.

Substance class and toxins	Retention time	Precursor ion	Adduct	Main product ion		Secondary product ion		Capillary Voltage
	[min]	[*m/z*]		[*m/z*]	Collision energy [eV]	[*m/z*]	Collision energy [eV]	[V]
Aflatoxins								
AFB_1	7.5	313.0	$[M+H]^+$	285.0	20	270.0	30	65
AFB_2	6.4	315.0	$[M+H]^+$	287.1	20	259.1	30	65
AFG_1	5.4	329.0	$[M+H]^+$	243.0	28	214.7	30	60
AFG_2	4.7	331.0	$[M+H]^+$	313.1	20	245.1	30	65
AFM_1	4.6	329.0	$[M+H]^+$	273.0	20	259.0	20	65
Alternaria toxins								
ALT	14.5	291.0	$[M-H]^-$	247.9	20	275.8	20	60
AME	19.4	271.1	$[M-H]^-$	255.7	20	227.6	20	60
AOH	14.6	257.0	$[M-H]^-$	212.8	20	214.8	20	70
Tet	16.3	413.2	$[M-H]^-$	271.0	10	140.9	10	65
Ergot alkaloids								
ECO	8.2/11.0	562.3	$[M+H]^+$	544.0	10	305.2	30	40
ECY	9.6/12.3	576.3	$[M+H]^+$	558.4	10	305.2	30	50
"Others"								
BEA	19.8	784.4	$[M+H]^+$	244.0	20	262.4	26	70
STC	15.0	325.0	$[M+H]^+$	310.1	20	281.3	40	35
SUL	6.9	333.2	$[M+H]^+$	208.9	10	136.2	40	25
Penicilliums								
CIT	16.2	281.1	$[M+Ac]^-$	204.8	20	248.0	20	35
PAT	3.0	152.9	$[M-H]^-$	108.9	10	80.9	10	30
RALs								
α-ZOL	17.9	319.4	$[M-H]^-$	275.0	20	160.0	30	60
β-ZOL	16.8	319.1	$[M-H]^-$	275.2	20	301.1	20	60
ZEA	18.2	317.4	$[M-H]^-$	175.0	20	131.0	25	60
Trichothecenes A								
DAS	7.2	384.3	$[M+NH_4]$	307.0	10	247.0	12	30
HT-2	10.3	442.2	$[M+NH_4]$	263.1	12	215.2	12	30
NEO	3.1	400.2	$[M+NH_4]$	305.0	11	185.0	20	20
T-2	12.4	484.4	$[M+NH_4]$	185.1	21	215.0	17	35
VERA	12.4	520.1	$[M+NH_4]$	249.1	20	231.2	20	45
Trichothecenes B								
3-AcDON	11.8	397.1	$[M+Ac]^-$	307.1	18	337.1	14	25
DON	5.1	355.1	$[M+Ac]^-$	295.4	10	264.8	6	50
FUS-X	9.2	413.1	$[M+Ac]^-$	353.0	14	263.0	11	35
NIV	2.7	371.1	$[M+Ac]^-$	280.9	10	311.0	6	35

3.2.5 Method optimization parameter

During the method optimization different extraction procedures, solvent mixtures, and clean-up cartridges were tested and evaluated. The extraction or clean-up recovery is defined by dividing the quantified amount of the analyte minus the native amount by the amount added before extraction, or clean-up (expressed in percentage). Six representatives (i.e. AFB_1, BEA, FUS-X, PAT, T-2 and ZEA) for the 28 mycotoxins included in the method were tested this way. They were selected due to their widely varying Log K_{ow} values from -0.13 (PAT) to 5.9 (BEA) (for details see[34]).

3.2.6 Method validation parameter

Linearity: The instrument linearity was tested with LC-MS/MS analysis of mycotoxins in Milli-Q water/MeOH (90:10, v/v) at concentrations between 0.1 ng/L and 150 x 10^3 ng/L. The linearity of the method was determined as the R^2 of the regression line obtained from the individual matrix-matched calibration solutions, once during the method validation, where the concentration ranged from 0 to 150 ng/g, and additionally during the method application, where concentrations ranged from 25 to 1000 ng/g.

Ion suppression: Matrix effects during analyte ionization causing suppression or enhancement of the analyte signal were evaluated by comparing the obtained signal from injection of the same amount of analyte in Milli-Q water/MeOH (90:10, v/v) (standard calibration curve) and an extracted matrix-matched sample. The signal suppression or enhancement, expressed in percentage (SSE) was quantified as 1 minus the ratio between the slope of the curve obtained for the matrix-matched extracts and the slope of the curve for the standard calibration curve as described by Gosetti et al.[35] The use of matrix-matched calibrations (see above) compensated for any losses or gains due to SSE.

Method recoveries, repeatability and reproducibility, instrument repeatability, blank concentrations and limits of detection/quantification: Wheat plant samples were spiked prior to solid-liquid extraction to produce concentrations of 50, 100 and 150 ng/g of each analyte. Five replicates for each concentration were generated. A non-spiked identical sample was analyzed in parallel with three replicates. The method recovery for each of the three spiking levels was determined by dividing the quantified amount of the analyte minus the native amount by the added amount.

The repeatability was determined as the relative standard deviation (RSD_r-M) of five concomitantly analyzed replicates at the three different concentrations mentioned above. Similarly, the instrument repeatability (RSD_r-I) was obtained from five consecutive analyses of an individual sample at a concentration of 100 ng/g. In addition the reproducibility was quantified for the natively present mycotoxins as relative standard deviation (RSD_R-M) of a real sample gathered on July 14, 2011, and analyzed three times over a period of eight weeks.

Limits of detection (LODs) and limits of quantification (LOQs) were calculated based on signal-to-noise (S/N) ratios of 3:1, and 10:1, respectively, as obtained from chromatograms of wheat plant extracts used for method validation. They either contained mycotoxins natively (in the case of 3-AcDON, DON, NIV, and ZEA, with concentrations of 2.6×10^2, 3.1×10^3, $3.5\ 10^2$, and 1×10^3 ng/g_{dw}, respectively), or were spiked with 50 ng/g. Note that LODs and LOQs obtained from high concentrations may be somewhat underestimated.

Due to the difficulty to find non-contaminated sample matrix for method validation and the fact that no reference material for whole above-ground wheat plants was available, a non-spiked native wheat sample was processed to assess the initial mycotoxin concentration and to obtain the optimal concentration range. If initial mycotoxin concentrations were higher than 150 ng/g, the sample extract was diluted to achieve the desired working range and was re-analyzed.

3.3. Results and Discussion

3.3.1 Optimization of the extraction and clean-up procedure

The entire extraction procedure was optimized with respect to the use of different extraction methods (SLE and ASE) to achieve the highest possible extraction recoveries. Although ASE has proven to be a reliable, and accurate method for the extraction of mycotoxins from crude matrices,[23, 27] however an increased amount of matrix components is usually extracted as well. With an initial solvent mixture of MeOH/MeCN (50:50, v/v), as mentioned by Urraca et al.,[23] the average extraction recovery for six test compounds (AFB_1, BEA, FUS-X, PAT, T-2 and ZEA) was 81% with SLE, but only 60% when ASE was used. Specifically, higher extraction recoveries and less interfering matrix components were observed for AFB_1, ZEA and the two trichothecenes, when a simple SLE approach was used. For details see **Error! Reference source not found.**.

Table 3.2. Overview of the recoveries [%] obtained for all different optimization steps during the method development: choice of the extraction procedure, the extraction solvent, as well as the clean-up procedure.

	Extraction		**Extraction solvent**			**Extraction clean-up**		
compound	SLE	ASE	ACN/MilliQ (80:20)	ACN/MilliQ/CH_3COOH (79:20:1)	MeCN/Milli-Q water/acetone (50:25:25)	Bond Elut Mycotoxin®	Trichothecene EP®	MycoSep Trichothecene®
AFB_1	88	49	85	82	37	79	10	nd
BEA	49	47	85	59	82	83	nd	nd
FUS-X	79	70	82	72	61	89	100	58
PAT	88	88	91	98	74	58	66	7
T-2	103	65	89	104	88	99	75	19
ZEA	71	37	88	70	62	98	24	79
Average	81	60	87	80	67	82	57	44

nd: not determined.

Generally, a crucial parameter for the development of any analytical method is the composition of the solvent applied for the extraction, particularly when dealing with compounds of high chemical diversity. Different mixtures of organic solvents (MeCN and MeOH) and water with and without the addition of acetic acid are frequently used for the extraction of various mycotoxins in cereals and related food products, as well as in silage and feed products.[30-32,36-38] Therefore, three different initial solvent mixtures, MeCN/Milli-Q water (80:20, v/v), MeCN/Milli-Q water/acetic acid (79:20:1, v/v/v), and MeCN/Milli-Q water/acetone (50:25:25, v/v/v), were tested for their extraction recovery. Extraction recoveries of six different test compounds (AFB_1, BEA, FUS-X, PAT, T-2 and ZEA) with these three solvent combinations were on average 87%, 80% and 67%, respectively (for details see **Table 3.2**). Hence, better results were obtained when acetone was omitted. The best recoveries were obtained with MeCN/Milli-Q water (80:20, v/v), although certain individual mycotoxins such as DON and ZEA may be even better extracted in the additional presence of acetone.[33] The addition of 1% acetic acid is further commonly advised by several authors[31,39] in order to obtain better recoveries for PAT or FB_1 and FB_2. However, due to the overall extraction recoveries obtained in our preliminary tests, we chose the solvent mixture MeCN/Milli-Q water (80:20, v/v) as most suitable and it was therefore further used for the optimization of the clean-up step.

Due to the complexity of the sample matrix and the aim to minimally pollute the LC-MS/MS equipment, three different clean-up cartridges, i.e., Trichothecene EP®, MycoSep Trichothecene® and Bond Elut Mycotoxin®, were tested for their analyte recovery and for their purification efficiency. All three materials were recommended for extract clean-up in food and feed matrices. Especially, Bond Elut Mycotoxin® was chosen due to its proprietary silica-based ion exchange material, which effectively retains a large number of compounds, basic, neutral and acidic.[40] Among the three, Bond Elut Mycotoxin® exhibited the best performance for the measured analytes with an average clean-up recovery of 82% at a spike concentration of 100 ng/g, compared to 57 and 44%, respectively (for details see **Table 3.2**). A significant reduction on SSE for all six test compounds from an average of 61% without, to 43% with the clean-up step. The corresponding increase in sensitivity is visualized in **Figure 3.2A** and **B**, where DON, T-2 and ZEA exhibit a 3, 1.5 and 12 times higher intensity, respectively, due to extract clean-up.

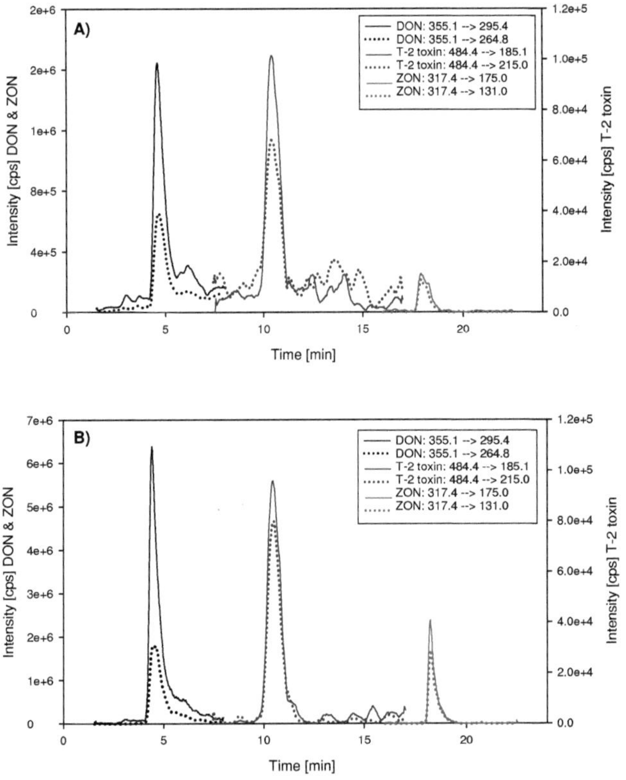

Figure 3.2. Overlay of three different LC-MS/MS ion chromatograms from non-spiked whole wheat plant extracts measured in positive and negative ionisation mode A) after solid-liquid extraction only and B) with an additional clean-up step using Bond Elut Mycotoxin®. Note the right y-axis for T-2 toxin. For details on the instrumental parameters, see text and Table 1.

3.3.2 Validation of the optimized method

Linearity: The MS/MS was linear for aqueous solutions between 0.1 ng/L and 150 x 10^3 ng/L, corresponding to 5 pg and 7.5 µg at the detector, respectively ($0.9410 \leq R^2 \leq 0.9998$). With $0.976 \leq R^2 \leq 1.000$, the matrix-matched calibration curves were linear over the whole calibration range from 25 to 1000 ng/g (**Table 3.3**).

Table 3.3. Figures of merits of the validated method: recovery (R), and repeatability (RSD_r-M) at three different concentrations, instrument repeatability (RSD_r-I), reproducibility (RSD_R-M), limits of detection and quantification (LOD and LOQ), signal suppression/enhancement (SSE), and linearity (R^2).

mycotoxin	50 ng/g		100 ng/g		150 ng/g		RSD_r-I	RSD_R-M	LOD[a]	LOQ[a]	SSE	R^{2b}
	R [%]	RSD_r-M [%]	R [%]	RSD_r-M [%]	R [%]	RSD_r-M [%]	[%]	[%]	[ng/g]	[ng/g]	[%]	
Aflatoxins												
AFB_1	69	2	88	3	82	10	3	nd	7	23	90	0.9981
AFB_2	90	7	96	6	90	8	1	nd	4	14	88	0.9968
AFG_1	73	4	74	10	71	9	2	nd	19	62	96	0.9988
AFG_2	101	27	98	8	100	8	3	nd	26	87	97	0.9991
AFM_1	92	10	93	5	89	12	3	nd	20	66	94	0.9979
Alternaria toxins												
ALT	107	2	101	4	99	1	8	nd	10	32	51	0.9988
AME	101	10	100	6	94	6	2	nd	1	3	31	0.9931
AOH	19	9	8	57	15	18	11	nd	20	68	41	0.9759
Tet	107	3	112	2	102	6	3	nd	3	9	83	0.99
Ergot alkaloids												
ECO	107	9	101	6	92	8	4	nd	9	29	71	0.9942
ECY	104	9	104	5	94	7	1	nd	9	29	66	0.9978
Others												
BEA	115	6	110	6	98	10	3	27	2	7	-89	0.9915
STC	99	6	98	2	91	6	4	nd	1	2	-10	0.9992
SUL	103	6	93	8	91	6	4	nd	14	45	84	0.9988
Penicilliums												
CIT	70	2	69	3	69	2	3	nd	1	2	7	0.991
PAT	99	4	81	3	76	2	9	nd	1	4	65	0.9966
Resorcyclic acid lactones												
α-ZOL	99	10	99	5	90	5	3	nd	2	6	70	0.9982
β-ZOL	122	8	112	4	101	2	9	nd	1	4	42	0.9999
ZEA	98	21	100	9	96	5	4	15	1	4	61	0.999

mycotoxin	50 ng/g		100 ng/g		150 ng/g		RSD_r-I	RSD_R-M	LOD[a]	LOQ[a]	SSE	R^{2b}
	R [%]	RSD_r-M [%]	R [%]	RSD_r-M [%]	R [%]	RSD_r-M [%]	[%]	[%]	[ng/g]	[ng/g]	[%]	
Trichothecenes A												
DAS	102	3	104	3	99	8	1	nd	1	3	61	0.998
HT-2	104	6	101	3	98	5	5	13	2	8	80	0.9998
NEO	82	4	78	3	76	8	2	nd	5	17	77	0.9998
T-2	104	1	103	2	96	5	2	4	2	7	43	0.9978
VERA	103	7	106	5	90	5	5	nd	5	17	49	1
Trichothecenes B												
3-AcDON	103	2	103	3	96	6	6	8	4	12	53	0.9992
DON	98	10	101	2	101	2	6	4	1	4	77	0.9999
FUS-X	102	4	102	2	100	3	2	13	8	26	64	0.995
NIV	101	4	100	4	90	10	9	2	8	25	84	0.9904
Median	101	6	100	4	94	6	3	10	4	14	61	0.9982
Min.	19	1	8	2	15	1	1	2	1	2	-89	0.9759
Max.	122	27	112	57	102	18	28	27	26	87	97	1

[a] Determined using a spiked wheat sample of 50 ng/g; [b] in matrix for the concentration range of 25 to 1000 ng/g; [c] Determined using a spiked wheat sample of 100 ng/g; nd: not detectable.

Ion suppression: Most of the mycotoxins were influenced by matrix effects (**Table 3.3**), with the AFs, TET, SUL, HT-2 and NIV being the most affected (SSE ≥ 80%). In contrast, BEA and STC showed strong enhancement (up to -89%). Such considerable high SSE once again underlines the need to quantitate mycotoxins by means of matrix-matched calibration curves or standard additions to compensate for drastic matrix effects.

Method recoveries: Over all three concentrations (50, 100, and 150 ng/g), very good recoveries of 69 to 122% (median 96%; excluding AOH) were obtained for all mycotoxins (**Table 3.3**), except for AOH. The low recoveries of AOH (8-19%) are probably caused by the lack of acetic acid, as Sulyok et al.[31] obtained satisfactory data in its presence. Therefore, the method should not be used for quantitative purposes for this compound. However, in the current study, the recoveries of several other compounds, like BEA and ZEA, were lower with acetic acid, and therefore further method optimization was done with the solvent introduced above. Overall, recoveries in the range of 70 to 120°% are acceptable according to good laboratory practice[41] and compromises with respect to performance for some individual analytes are in the nature of multi-residue methods. In comparison with other multi-mycotoxin methods,[32] we achieved comparable, or in the case of selected *Fusarium* toxins like BEA and DON, even better recoveries.

Repeatability and reproducibility, instrument repeatability, blank concentrations and limits of detection/quantification: Method repeatability (RSD_r-M) ranged from 1 to 18% at 150 ng/g, with a median value of 6% (**Table 3.3**). At 100 ng/g, a median precision level of 4% was determined, ranging from 2 to 57%. Here, the worst precision was observed for AOH possibly caused by the very low overall recovery of just 8%. At 50 ng/g, RSD_r-M ranged from 1 to 27% (median 6%; **Table 3.3**). Here, the outlier with just 27% precision was AFG_2. The overall RSD_r-M over all three concentrations showed a median of 6%, with no discernible concentration dependence. Considering the structural diversity of the analytes, the precision obtained for this multi-component method is satisfactory[31, 42] and confirms the robustness of the analytical method. In comparison, the instrument repeatability (RSD_r-I) was between 1 and 28% for all mycotoxins in the same matrix at a concentration of 100 ng/g (**Table 3.3**). This indicates that most of the uncertainty in repeatability was caused by sample analysis and not by sample preparation. The relative standard deviation of the real sample triplicate, here referred to as reproducibility (RSD_R-M; **Table 3.3**), ranged from 2% for NIV to 27% for BEA. As expected, these percentages were similar (3-AcDON, DON, and T-2) or slightly higher (BEA, HT-2, FUS-X, and ZEA) than the RSD_r-M.

The LODs ranged from 1 to 26 ng/g_{dw}, with a median of 4 $ng/g_{dw.}$ Although no maximum limits or recommendations of maximum allowed concentrations are specifically set for mycotoxins in whole wheat plants, the LODs obtained here can be compared with earlier methods investigating mycotoxin occurrence in maize, grass and wheat silage. For example, Garon et al.[37] reported LODs of 6.5 ng/g for ZEA and DON in maize silage. These values are three times higher than those presented here (LOD of 1 ng/g_{dw} for both compounds), even though our method comprises multiple classes of mycotoxins. Driehuis et al.[43] reported much higher LOQs than those of the current study, especially for the trichothecenes DON, 3-AcDON, HT-2 and T-2, as well as ZEA. The method presented here performed also better compared with a recently established multi-mycotoxin method, where 26 different mycotoxin were quantified in maize silage.[32] Except for the AFs, where the method from Van Pamel et al.[32] resulted in clearly lower LODs (5-7 ng/g, compared with 7-26 ng/g in our method), the here introduced method was superior, especially for trichothecenes (e.g. DON LOD of 50 ng/g in maize silage, and LOD of 1 ng/g in whole wheat plants), as well as for BEA, CIT, and the RALs. LODs obtained here are comparable with those published for trichothecenes B and RALs,[44] or even better in comparison with the application of a multi-mycotoxin method for the analysis of different corn flour samples.[26]

3.3.3 Method application

The validated method is currently applied to examine the composition and abundance of mycotoxins of four different *Fusarium* species in whole wheat plants from the experimental field at our research station.

Initial results from this field study for the part, which was artificially infested with *F. crookwellense*, are summarized in **Table 3.4**. If quantified concentrations exceeded the range of the matrix-matched calibration (i.e. were higher than 1000 ng/g), the extracts were diluted and re-measured. Generally, the concentrations obtained with the diluted extracts were comparable to those obtained with the original sample (0 - 8% higher). In 2010, NIV showed the highest concentration of all analyzed mycotoxins with 10 $\mu g/g_{dw}$, whereas DON was quantified with up to 3 $\mu g/g_{dw}$, followed by ZEA with up to 2 $\mu g/g_{dw}$. Additionally, BEA and 3-AcDON were quantified as well, but at lower concentrations. Comparable concentrations of NIV and ZEA were found in artificially infected winter wheat grains,[45] whereas lower concentrations were reported for naturally contaminated winter wheat grains.[46] Overall, the mycotoxin pattern presented here is in good agreement with the mycotoxins commonly associated with *F. crookwellense*.[47,48] Usually, NIV is the most abundant mycotoxin produced by *F. crookwellense*, which is in accordance with the results of this study.

In 2011, a highly dynamic mycotoxin-occurrence pattern was discernible over the four different sampling times within the growing season (**Table 3.4**). On May 26, 2011, the wheat was artificially infected and three weeks after, the first plant samples were taken from the test field. Except for NIV, HT-2 and T-2, no other mycotoxins were detected at this early stage of fungal development. Interestingly, the highest concentration of NIV was detected five weeks after the artificial infection, at the end of June 2011. Afterwards and until harvest, the concentration of NIV decreased, but was still twice as high as the concentration of DON. This decrease in the NIV concentration has – to our knowledge – not been reported before, but is in analogy with results documented in three different studies for another type B trichothecene, namely DON.[13,49,50] These authors observed peak concentrations of DON in wheat grains prior to harvest and attributed the rapid decline in the DON concentration in the final phase of wheat maturation to the breakdown of this compound by plant enzymes or a conjugation of DON to glucosides.[51,52] Similarly, NIV may form glucosides as well.[53] Additionally, Rohweder et al.[50] stated that elution of water soluble DON due to rainfall might be another potential explanation for the decline in DON concentration of wheat plants. This underlines the possibility of mycotoxins entering also the aquatic environment, due to wash-out of mycotoxins from infected plants via run-off or drainage from the field, as illustrated by several authors in the recent past.[14,18,21] For DON and BEA, peak concentrations were detected at the time of harvest. Their unexpected high concentrations in the *F. crookwellense* plot were probably the result from wind dispersed conidia from wheat straw of neighboring plots infected with *F. graminearum* and *F. avenaceum* or from drift of spore suspensions during the artificial infection of these plots. A complete overview about the results including the mycotoxin concentration pattern of all parts will be published elsewhere.

Table 3.4. Concentrations [ng/gdw] of the mycotoxins found in F. crookwellense infected whole wheat plants

	Type A trichothecenes		Type B trichothecenes					
Sampling date	HT-2	T-2	3-AcDON	DON	FUS-X	NIV	BEA	ZEA
29.07.2010	*nd*	*nd*	*168*	*3159*[c]	*nd*	*9824*[c]	*32*	*1842*[c]
16.06.2011	516	34	nd	nd	nd	6503[c]	nd	nd
30.06.2011	305	131	nd	206	680	11891[c]	nd	91
14.07.2011	189 ± 25[b]	69 ± 3[b]	112 ± 9[b]	2053 ± 81[b,c]	242 ± 31[b]	3868 ± 79[b,c]	679 ± 180[b]	459 ± 70[b]
16.07.2011	*103*	*44*	*71*	*2681*[c]	*179*	*5575*[c]	*1782*[c]	*83*

[a] nd: not detected; [b] triplicate measurements ± standard deviation of three measurements; [c] Concentration outside the range

In conclusion, the developed method is the first to quantify 28 mycotoxins in above-ground material from whole wheat plants. The extraction procedure includes a user-friendly solid-liquid extraction step with a subsequent cartridge clean-up for the reduction of matrix components, followed by analysis with HPLC-ESI(+/-)-MS/MS. The use of the matrix-matched calibrations is recommended due to the lack of available blank samples and reference materials. Thereby, isotopically labeled internal standards could be omitted, which are not available for all compounds included in this method and/or are very expensive. The method was successfully applied to whole-wheat plant samples, in which quantifiable amounts of namely 3-AcDON, DON, FUS-X, HT-2, NIV, T-2, BEA and ZEA were detected. Under good agricultural practice (including plowing, a proper pest management, crop rotation, as well as the cultivation of less *Fusarium* susceptible wheat varieties), and without any artificial plant infestation, the mycotoxin concentrations should be usually much lower. Nevertheless, to establish a deeper understanding about the mycotoxin production and distribution within a wheat plant, it would be necessary to analyze the different plant parts separately (grains, glumes, spindles, and husks) during the different development phases from the flowering on.

Acknowledgment

We would like to thank Andreas Hecker for his help in setting up the field site, inoculating the plants and for sharing his insights on *Fusarium* diseases. Additionally, we acknowledge the work of the field staff at Agroscope ART and their great support at the field site.

Source of funding

We thank the Swiss Federal Office for the Environment for the financial support of this project.

Conflict of interest

None.

References (chapter 3)

1. Cole, R. J., et al., *Handbook of Secondary Fungal Metabolites.* Academic Press: New York: 2003; Vol. III.
2. Desjardin, A. E., *Fusarium Mycotoxins - Chemistry, Genetics and Biology.* 1. ed.; The American Phytopathological Society: St. Paul, Minnesota 55121, U.S.A., 2006; Vol. 1, p 240.
3. Trucksess, M. W., Mycotoxins. *Journal of Aoac International* **2001,** *84*, 202-211.
4. Lanier, C., et al., Mycoflora and mycotoxin production in oilseed cakes during farm storage. *J. Agric. Food Chem.* **2009,** *57*, 1640-1645.
5. Faberi, A., et al., Determination of type B fumonisin mycotoxins in maize and maize-based products by liquid chromatography/tandem mass spectrometry using a QqQ(linear ion trap) mass spectrometer. *Rap. Commun. Mass Spectrom.* **2005,** *19*, 275-282.
6. Zollner, P.; Mayer-Helm, B., Trace mycotoxin analysis in complex biological and food matrices by liquid chromatography-atmospheric pressure ionisation mass spectrometry. *J. Chromatogr. A* **2006,** *1136*, 123-169.
7. Krska, R., et al., Analysis of *Fusarium* toxins in feed. *Ani. Feed Sci. Technol.* **2007,** *137*, 241-264.
8. Coppock, R. W.; Jacobsen, B. J., Mycotoxins in animal and human patients. *Toxicol. Ind. Health* **2009,** *25*, 637-655.
9. Kuiper-Goodman, T., Food safety: Mycotoxins and phycotoxins in perspective. In *Mycotoxins and Phycotoxins - Developments in Chemistry, Toxicology and Food Safety*, Miraglia, M.; VanEgmond, H. P.; Brera, C.; Gilbert, J., Eds. Alaken, Inc: Ft Collins, 1998; pp 25-48.
10. Bennett, J. W., Mycotoxins, mycotoxicoses, mycotoxicology and mycopathologia. *Mycopathol.* **1987,** *100*, 3-5.
11. Hussein, H. S.; Brasel, J. M., Toxicity, metabolism, and impact of mycotoxins on humans and animals. *Toxicol.* **2001,** *167*, 101-134.

12. Richard, J. L., Some major mycotoxins and their mycotoxicoses - An overview. *Int. J. Food Microbiol.* **2007,** *119*, 3-10.

13. Brinkmeyer, U., et al., Influence of a *Fusarium culmorum* inoculation of wheat on the progression of mycotoxin accumulation, ingredient concentrations and ruminal in sacco dry matter degradation of wheat residues. *Arch. Ani. Nutri.* **2006,** *60*, 141-157.

14. Hartmann, N., et al., Occurrence of zearalenone on *Fusarium graminearum* infected wheat and maize fields in crop organs, soil and drainage water. *Environ. Sci. Technol.* **2008,** *42*, 5455-5460.

15. Vogelgsang, S., et al., On-farm experiments over 5 years in a grain maize/winter wheat rotation: effect of maize residue treatments on *Fusarium graminearum* infection and deoxynivalenol contamination in wheat. *Mycotoxin Res.* **2011,** *27*, 81-96.

16. Scudamore, K. A.; Livesey, C. T., Occurrence and significance of mycotoxins in forage crops and silage: a review. *J. Sci. Food Agric.* **1998,** *77*, 1-17.

17. Gutzwiller, A.; Gafner, J. L., Mycotoxin contaminated bedding straw and sow fertility. *Rev. Suisse Agric.* **2008,** *40*, 139-142.

18. Bucheli, T. D., et al., *Fusarium* mycotoxins: Overlooked aquatic micropollutants? *J. Agric. Food Chem.* **2008,** *56*, 1029-1034.

19. Hartmann, N., et al., Quantification of estrogenic mycotoxins at the ng/L level in aqueous environmental samples using deuterated internal standards. *J. Chromatogr. A* **2007,** *1138*, 132-140.

20. Hartmann, N., et al., Quantification of zearalenone in various solid agroenvironmental samples using D-6-zearalenone as the internal standard. *J. Agric. Food Chem.* **2008,** *56*, 2926-2932.

21. Schenzel, J., et al., A multi residue screening method to quantify mycotoxins in aqueous environmental samples. *J. Agric. Food Chem.* **2010,** *58*, 11207-11217.

22. Vendl, O., et al., Simultaneous determination of deoxynivalenol, zearalenone, and their major masked metabolites in cereal-based food by LC-MS-MS. *Anal. Bioanal. Chem.* **2009,** *395*, 1347-1354.

23. Urraca, J. L., et al., Analysis for zearalenone and alpha-zearalenol in cereals and swine feed using accelerated solvent extraction and liquid chromatography with fluorescence detection. *Anal. Chim. Acta* **2004,** *524*, 175-183.

24. D'Arco, G., et al., Analysis of fumonisins B-1, B-2 and B-3 in corn-based baby food by pressurized liquid extraction and liquid chromatography/tandem mass spectrometry. *J. Chromatogr. A* **2008,** *1209*, 188-194.

25. Berthiller, F., et al., Rapid simultaneous determination of major type A- and B-trichothecenes as well as zearalenone in maize by high performance liquid chromatography–tandem mass spectrometry. *J. Chromatogr. A* **2005,** *1062*, 209 - 216.

26. Cavaliere, C., et al., Development of a multiresidue method for analysis of major Fusarium mycotoxins in corn meal using liquid chromatography/tandem mass spectrometry. *Rap. Commun. Mass Spectrom.* **2005,** *19*, 2085-2093.

27. Desmarchelier, A., et al., Development and Comparison of Two Multiresidue Methods for the Analysis of 17 Mycotoxins in Cereals by Liquid Chromatography Electrospray Ionization Tandem Mass Spectrometry. *J. Agric. Food Chem.* **2010,** *58*, 7510-7519.

28. Jin, P. G., et al., Simultaneous determination of 10 mycotoxins in grain by ultra-high-performance liquid chromatography-tandem mass spectrometry using 13C15-deoxynivalenol as internal standard. *Food Add. Contam. Part A-Chem. Anal. Control Exp. Risk Assess.* **2010,** *27*, 1701-1713.

29. Cavaliere, C., et al., Mycotoxins produced by *Fusarium* genus in maize: determination by screening and confirmatory methods based on liquid chromatography tandem mass spectrometry. *Food Chem.* **2007,** *105*, 700-710.

30. Rasmussen, R. R., et al., Multi-mycotoxin analysis of maize silage by LC-MS/MS. *Anal. Bioanal. Chem.* **2010,** *397*, 765-776.

31. Sulyok, M., et al., Development and validation of a liquid chromatography/tandem mass spectrometric method for the determination of 39 mycotoxins in wheat and maize. *Rap. Commun. Mass Spectrom.* **2006,** *20*, 2649-2659.

32. Van Pamel, E., et al., Ultrahigh-performance liquid chromatographic tandem mass spectrometric multimycotoxin method for quantitating 26 mycotoxins in maize silage. *J. Agric. Food Chem.* **2011,** *59*, 9747-9755.

33. Dorn, B., et al., *Fusarium* species complex and mycotoxins in grain maize from maize hybrid trials and from grower's fields. *J. Appl. Microbiol.* **2011,** *111*, 693-706.

34. Schenzel, J., et al., Experimentally determined soil organic matter-water partitioning coefficients for different classes of natural toxins and comparison with estimated numbers. **in prep.**

35. Gosetti, F., et al., Signal suppression/enhancement in high-performance liquid chromatography tandem mass spectrometry. *J. Chromatogr. A* **2010,** *1217*, 3929-3937.

36. Frenich, A. G., et al., Simple and high-throughput method for the multimycotoxin analysis in cereals and related foods by ultra-high performance liquid chromatography/tandem mass spectrometry. *Food Chem.* **2009,** *117*, 705-712.

37. Garon, D., et al., Mycoflora and multimycotoxin detection in corn silage: Experimental study. *J. Agric. Food Chem.* **2006,** *54*, 3479-3484.

38. Monbaliu, S., et al., Development of a multi-mycotoxin liquid chromatography/tandem mass spectrometry method for sweet pepper analysis. *Rap. Commun. Mass Spectrom.* **2009,** *23*, 3-11.

39. Krska, R., et al. In *Mycotoxin analysis: An update*, 12th IUPAC International Symposium on Mycotoxins and Phycotoxins, Istanbul, TURKEY, May 21-25, 2007; Taylor & Francis Ltd: Istanbul, TURKEY, 2007; pp 152-163.

40. Klotzel, M., et al., A new solid phase extraction clean-up method for the determination of 12 type A and B trichothecenes in cereals and cereal-based food by LC-MS/MS. *Molec. Nutri. Food Res.* **2006,** *50*, 261-269.

41. Jiang, W., Good Laboratory Practice in Analytical Laboratory. *The Journal of American Science* **2005,** *1*, 93-94.

42. Capriotti, A. L., et al., Development and validation of a liquid chromatography/atmospheric pressure photoionization-tandem mass spectrometric method for the analysis of mycotoxins subjected to commission regulation (EC) No. 1881/2006 In cereals. *J. Chromatogr. A* **2010,** *1217*, 6044-6051.

43. Driehuis, F., et al., Occurrence of mycotoxins in feedstuffs of dairy cows and estimation of total dietary intakes. *J. Dairy Sci.* **2008,** *91*, 4261-4271.

44. Di Mavungu, J. D., et al., LC-MS/MS multi-analyte method for mycotoxin determination in food supplements. *Food Add. Contam. Part A-Chem. Anal. Control Exp. Risk Assess.* **2009,** *26*, 885-895.

45. Christ, D. S., et al., Pathogenicity, symptom development, and mycotoxin formation in wheat by *Fusarium* species frequently isolated from sugar beet. *Phytopathol.* **2011,** *101*, 1338-1345.

46. Edwards, S. G., *Fusarium* mycotoxin content of UK organic and conventional wheat. *Food Add. Contam. Part A-Chem. Anal. Control Exp. Risk Assess.* **2009,** *26*, 496-506.

47. Buck, H. T., et al., *Wheat production in stressed environments.* Springer: Dordrecht, The Netherlands, 2007; Vol. 1.

48. Bottalico, A.; Perrone, G., Toxigenic *Fusarium* species and mycotoxins associated with head blight in small-grain cereals in Europe. *Eur. J. Plant Pathol.* **2002,** *108*, 611-624.

49. Matthaus, K., et al., Progression of mycotoxin and nutrient concentrations in wheat after inoculation with Fusarium culmorum. *Arch. Ani. Nutri.* **2004,** *58*, 19-35.

50. Rohweder, D., et al., Effect of different storage conditions on the mycotoxin contamination of *Fusarium culmorum*-infected and non-infected wheat straw. *Mycotoxin Res.* **2011,** *27*, 145-153.

51. Engelhardt, G., et al., Metabolism of mycotoxins in plants. *Adv. Food Sci.* **1999,** *21*, 71-78.

52. Berthiller, F., et al., Masked mycotoxins: Determination of a deoxynivalenol glucoside in artificially and naturally contaminated wheat by liquid chromatography-tandem mass spectrometry. *J. Agric. Food Chem.* **2005,** *53*, 3421-3425.

53. Nakagawa, H., et al., Detection of a new *Fusarium* masked mycotoxin in wheat grain by high-resolution LC-Orbitrap (TM) MS. *Food Add. Contam. Part A-Chem. Anal. Control Exp. Risk Assess.* **2011,** *28*, 1447-1456.

CHAPTER 4:

EXPERIMENTALLY DETERMINED SOIL ORGANIC MATTER-WATER SORPTION COEFFICIENTS FOR DIFFERENT CLASSES OF NATURAL TOXINS AND COMPARISON WITH ESTIMATED NUMBERS

JUDITH SCHENZEL,[1,2] KAI-UWE GOSS,[3] RENÉ P. SCHWARZENBACH,[2] THOMAS D. BUCHELI,[1]
STEVEN T. J. DROGE[3]

[1]AGROSCOPE RECKENHOLZ-TAENIKON, RESEARCH STATION ART, RECKENHOLZSTRASSE 191, CH-8046 ZURICH, SWITZERLAND

[2] INSTITUTE OF BIOGEOCHEMISTRY AND POLLUTANT DYNAMICS, SWISS FEDERAL INSTITUTE OF TECHNOLOGY, RAEMISTRASSE 101, CH-8092 ZURICH, SWITZERLAND

[3] DEPARTMENT OF ANALYTICAL ENVRIONMENTAL CHEMISTRY, HELMHOLTZ CENTRE FOR ENVIRONMENTAL RESEARCH UFZ, PERMOSERSTRASSE 15, D-04318 LEIPZIG, GERMANY

REPRODUCED WITH PERMISSION FROM:

ENVIRONMENTAL SCIENCE AND TECHNOLOGY, 2012

VOLUME 46, ISSUE 11, PAGES 6118-6126

Abstract

Although natural toxins such as mycotoxins or phytoestrogens are widely studied and were recently identified as micropollutants in the environment, many of their environmentally relevant physico-chemical properties have not yet been determined. Here, the sorption affinity to Pahokee peat, a model sorbent for soil organic matter, was investigated for 29 mycotoxins and two phytoestrogens. Sorption coefficients (K_{oc}) were determined with a dynamic HPLC-based column method using a fully aqueous mobile phase with 5 mM $CaCl_2$ at pH 4.5. Sorption coefficients varied from less than $10^{0.7}$ L/kg$_{oc}$ (e.g. all type B trichothecenes) to $10^{4.0}$ L/kg$_{oc}$ (positively charged ergot alkaloids). For the neutral compounds the experimental sorption data set was compared with predicted sorption coefficients using various models, based on molecular fragment approaches (EPISuite's KOCWIN or SPARC), poly-parameter linear free energy relationship (pp-LFER) in combination with predicted descriptors, and quantum-chemical based software (*COSMOtherm*)). None of the available models was able to adequately predict absolute K_{oc} numbers and relative differences in sorption affinity for the whole set of neutral toxins, largely because mycotoxins exhibit highly complex structures. Hence, at present, for such compounds fast and consistent experimental techniques for determining sorption coefficients, as the one used in this study, are required.

4.1 Introduction

Fungi of the genera *Alternaria*, *Aspergillus*, *Claviceps*, *Fusarium*, and *Penicillium* infect agricultural commodities and food worldwide, and produce a plethora of toxic metabolites. Prominent examples of these so-called mycotoxins include aflatoxins, alternaria toxins, trichothecenes, resorcyclic acid lactones, and ergot alkaloids. They contaminate food, beverages and feed, and have been ranked as the most important chronic dietary risk factor, higher than synthetic contaminants, plant toxins, food additives or pesticide residues.[1] Estimated crop losses and human and animal health costs were up to billions of dollars per year worldwide.[2,3] Consequently, much research has been carried out on the analysis, exposure assessment, and health effects of mycotoxins.[2,4,5] Only recently, the environmental exposure to mycotoxins has been investigated in more detail. We demonstrated emission of zearalenone (occasional) and deoxynivalenol (frequent) into the aquatic environment via drainage and run-off from infested agricultural sites.[6-8] In addition, besides field run-off, human excretions may be another relevant source of mycotoxins in the aquatic environment,[9,10] depending upon the removal rate in wastewater treatment plants (WWTPs). Deoxynivalenol, for example, is only partially eliminated by Swiss WWTPs.[11] Trace levels of nivalenol, 3-acetyl-deoxynivalenol, and beauvericin were recently detected repeatedly in various aquatic environments.[12]

Next to production and degradation of mycotoxins, sorption is a key parameter for a comprehensive assessment of their environmental fate. Sorption controls transport processes such as run-off or leaching to groundwater, as well as bioavailability to exposed biota, including microbial degraders of such compounds. Generally, sorption of neutral organic compounds in soils is dominated by their distribution between natural organic matter (NOM) and the surrounding pore water phase.[13] To date, measured NOM-water sorption coefficients (K_{oc}) are not determined for mycotoxins from an environmental fate point of view, except for the estrogenic compound zearalenone.[14] Most sorption studies involving mycotoxins have focused on feed additive sorbents in order to reduce mycotoxin exposure in farm animals. Humic material has also been considered as mycotoxin binding additive, but so far only for aflatoxin B_1,[15] deoxynivalenol,[16] ochratoxin A,[17] and zearalenone.[16,17] The main goal of this work was to establish a consistent set of experimentally determined sorption coefficients to soil organic matter for a diverse set of mycotoxins and two phytoestrogens.

Soil-sorption coefficients are typically determined in batch experiments such as OECD guideline 106, but can also be performed for moderately strong sorbing compounds in dynamic HPLC-based methods, using custom packed-columns with natural soils or purified soil components and aqueous phase eluents.[18-20] This HPLC column method to derive K_{oc} values has recently been extensively evaluated, and demonstrated (i) to agree well with batch sorption measurements for a series of alkylbenzenes,[21] indicating that local equilibrium in the column is attained for compounds of different hydrophobicity, (ii) to give consistent data when measured during several months and between different column packings,[21,22] and (iii) that Pahokee peat, a rather well standardized reference material, functions as a good representative of the sorption properties of NOM in a wide variety of soils.[20] We refer to references[20-22] for a more detailed account of the column method validation and for using Pahokee peat as model NOM material. For the current work, the sorption coefficients of 31 natural toxins, including several classes of mycotoxins and two phytoestrogens, to Pahokee peat were measured using a dynamic HPLC-based flow-through method. Besides verifying analytical consistency, sorption nonlinearity was specifically addressed in order to facilitate comparison between compounds. The reported sorption coefficients in this study form to our knowledge the first consistent data set for mycotoxins.

In the absence of experimental sorption data, a variety of predictive models and tools can be used to estimate NOM sorption coefficients. The second goal of this study was to compare the experimentally derived sorption data with various predictive models that are commonly used. This is particularly interesting for this set of natural toxins, because it comprises a group of compounds with a broad diversity of multifunctional groups that may not be included in calibration data sets for some predictive models. In addition, several (myco)toxins are negatively or positively charged under field conditions. Charged organic compounds have also been demonstrated to sorb to NOM,[23-25] but most existing sorption models only account for the fraction of neutral chemical species and may therefore deviate from measured NOM sorption coefficients. Because for apolar organic compounds sorption to NOM is largely a hydrophobic effect, for such compounds, NOM/water sorption coefficients have been found to correlate well with the corresponding octanol-water partition coefficients (K_{ow}).[26] Although octanol is not always a representative phase for NOM in the case of polar organic compounds,[13] K_{ow} based models are still the most common way to estimate the NOM sorption affinity. Unfortunately, there are no experimental K_{ow} values for any of the tested (myco)toxins, except for the two phytoestrogens.[27] As a result, if risk assessment models based on K_{ow} are to be used, also these need to be derived from predictive models, and therefore several models are used to estimated K_{ow} values in this study for natural toxins.

Using the measured K_{oc} data for the neutral mycotoxins, we evaluated the performance of five different approaches to estimate NOM sorption coefficients to deal with such complex structures: (i) a single parameter approach using (estimated) K_{ow} values,[26] (ii) a method based on additive contributions of molecular fragments calibrated on soil sorption data (EPISuite's KOCWIN),[28] (iii) a method based on a unified partitioning approach using various molecular descriptors and different NOM input structures for the sorbent phase (SPARC),[29] (iv) a method based on polyparameter linear free energy relationships (pp-LFERs) calibrated for soils or organic matter,[20,30,31] and (v) a quantum-chemical based software approach (*COSMOtherm*), that does not depend on a specific calibration but that applies statistical thermodynamics using only molecular structure of the solutes and sorbent/solvent as input information.[32-34]

4.2 Materials and Methods

4.2.1 Sorbents and solutes

Pahokee Peat (PPII 2S103P) was obtained from the International Humic Substances Society (IHSS, Golden, CO, USA). Properties and elemental composition are described in the Supporting Information (SI) in **Table S4.1**. The same batch of micronized peat was used by Bronner and Goss,[21] in which no particles larger than 20 μm were detected. **Section SI-4.1** lists details on the tested natural toxins, with information on compound purity and supplier, and molecular structures (**SI-1 Figure S4.1**). Estimated pK_a (SPARC v4.5, http://archemcalc.com/sparc), detection wavelength and operational concentration range are presented in **Table 4.1**, details on preparation of stock solutions and dilutions in **SI Section SI-4.2**.

4.2.2 Sorption experiment set-up

The dynamic HPLC-column method measures retention of compounds on custom packed columns, filled with a dry mixture of micronized Pahokee Peat and the relatively inert packing material silicon carbide (SiC, diameter ~3 μm, ESK-SiC, Frechen, Germany). Several packed columns were made with peat mixed to SiC to broaden the applicable retention window; a 5 cm stainless steel (ss) column with 10:100 ratio of peat:SiC (w/w) facilitated measurements of low sorbing compounds, whereas 1 cm ss-columns of 5:100, 3:100, 1.5:100 and 1:100 (~0.1 g total weight of column material) were used to facilitate the retention measurements of stronger sorbing compounds. As a result, the packed amount of peat in the different columns varied by about a factor 50, from 0.41–20 mg_{oc}. Air in freshly packed columns was purged out under a low water flow at ~100 bar by placing a restrictor after the column. Columns were further conditioned to the mobile phase for 3 more days at 0.1 mL/min, after which pH of the eluents in the column outflow was checked spectrophotometrically using trithiocyanuric acid.[20,22] A comparison of the different columns and types of micronized peat is presented in the **Section SI-4.3**.

The mobile phase in the HPLC system consisted of MilliQ-water of pH 4.5 with 5 mM $CaCl_2$. Most natural soils have Ca^{2+} concentrations ranging between 1-15 mM,[35] and 5 mM Ca^{2+} was also used in recent column studies with peat.[21,22] Most test compounds have *pKa* values > 7. At pH 4.5, these compounds will be tested in neutral form and, probably, display their strongest sorption affinities to NOM. For citrinin, which is largely anionic at pH 4.5 and higher, and for the ergot alkaloids, which are largely cationic below pH 7, the obtained sorption coefficients at pH 4.5 will be indicative values for NOM in many environmental systems. Note that this does not fully extend to the use of sorption coefficients for feed additives, which range between pH 3-8 (stomach-intestines, respectively). The eluent was pumped (Jasco 890PU) through the column at a flow rate of 0.05 mL/min, which was previously demonstrated to achieve >98% equilibration of analytes with micronized peat in similar columns.[21] A Rheodyn injector was used to switch a 200 μL ss-loop with test solution in- or offline with the eluent flow. For compound peak detection a UV/Vis-detector (Jasco 870UV) was used. All experiments were conducted at 23 ± 3°C, and performed in two time periods to check the long-term stability of custom packed peat columns, as was also reported elsewhere.[21]

4.2.3 Determination of sorption coefficients

Organic carbon normalized sorption coefficients (K_{oc}) were derived, as described elsewhere,[20] from the final net retention volume of the compound *i* ($V_{net\ i\ final}$; in L), divided by the amount of organic carbon (m_{OC}, in kg) of Pahokee Peat (see **SI-4.1 Table S4.1**) present in the column:

$$K_{OC} = \frac{V_{net\ i\ final}}{m_{OC}} \quad (1).$$

All further needed retention volumes ($V_{SiC,tracer}$, $V_{SiC,i}$, $V_{Peat,tracer}$, and $V_{ret,i\ Peat}$) to determine $V_{net,i\ final}$ are described in **Section SI-4.3**. All volumes were determined from the volumetric flow rate of the mobile phase (mL/min) multiplied by the retention time (min) of the respective peaks, the latter obtained from the first statistical moment according to Valocchi,[36] and demonstrated in **Figure S4.2**. Sorption to the chromatographic system (capillaries, frits, column wall) and SiC packing material was negligible for most compounds in comparison with sorption to NOM in the column. The uncertainty on K_{oc} values increases considerably when net retention volumes ($V_{net,i\ final}$) become very small compared to those for the non-retaining tracer ($V_{Peat,tracer}$). Therefore, all data for which $V_{net,i\ final}$ was below 30% of $V_{Peat,tracer}$ were discarded, resulting in a lower limit of log K_{oc} ~0.7 for the retardation range of the dynamic HPLC method with the 5 cm 10:100 column.

Because sorption to peat can be a non-linear process, retention volumes can increase slightly with decreasing (sorbed) concentrations.[20] In order to compare the sorption affinity of all test compounds at a single test concentration, we addressed this nonlinearity by constructing sorption isotherms from the retention data. To this end, at least three different dilutions of each compound were tested on a column. For several compounds a few more data points were generated by (i) using different columns or (ii) to verify sorption nonlinearity for a broad concentration range on a single column. In order to create sorption isotherms, a representative dissolved concentration must be derived from each eluted peak. Multiplying this estimated dissolved concentration ($C_{AQ,est}$) with the measured K_{oc} provides the representative sorbed concentration ($C_{S,est}$), from which isotherms can be created. Three different ways to derive $C_{AQ,est}$ were assessed, as discussed in **Section SI-4.3** and **Figures S4.2 & S4.3**. All resulted in the same "semi-quantitative" sorption isotherms (though not similar $C_{AQ,est}$). Because the peak maximum of each UV-signal recording was easily defined, we used this to calculate $C_{AQ,est}$ (using a single external calibration point on a spectrophotometer). Assuming generic Freundlich isotherms to fit the data, the obtained data for different concentrations should fit to a linear line on a logarithmic isotherm plot.

Sorption nonlinearity is most likely the result of sorption site heterogeneity, and therefore the best way to compare the sorption affinity between compounds on the same NOM is based on equal sorbed concentrations. In this study, we normalized the sorption coefficients to 100 mg/kg_{oc} because for most compounds the isotherms were in or close (within a factor 3, see **Table S4.4**) to this range. For soils with an OC content of 1-10%, the K_{oc} at 100 mg/kg_{oc} translates to soil-water distribution coefficients (K_d) at soil concentrations of 1-10 mg/kg, respectively.

4.2.4 Predictive models

Several software tools were used to predict K_{ow} and K_{oc} for the neutral (myco)toxins. For details on the selected prediction tools, we refer to **Section SI-4.4** and references given therein. The advantages and drawbacks of these techniques are discussed in detail elsewhere.[31,37-41]

4.3 Results and discussion

4.3.1 Deriving K_{OC} values for natural toxins using the dynamic HPLC-column method

Table 4.1 lists all experimentally determined K_{oc} values with the Pahokee-Peat columns method. The K_{oc} values and structures of the natural toxins are depicted in **Figure 4.1**. Retention volumes determined on the column with the highest packed amount of peat (5 cm x 10:100 peat:SiC) were too low for all type B trichothecenes (see structures in **Figure S4.1**), and also too low for some of type A (**Figure 4.1**, verrucarol, diacetoxyscirpenol, neosolaniol). For the (myco)toxins biochanin A, coumestrol and beauvericin, retention on the pure SiC column was too high to determine net retention volumes on peat columns with high precision, and they were therefore excluded.

Using estimated aqueous and sorbed concentrations, sorption isotherms could be plotted (see **Figure S4.4-S4.5**) in an overall range of 0.5- 6800 mg/kg_{oc} for all compounds (see further details in **Table S4.4**). For most compounds, Freundlich isotherms with exponents 0.7-1.1 fitted the log-based sorption data with high correlation coefficients. The close fits of these exponential sorption curves demonstrated that sorption affinity can be derived with fairly high precision at different concentrations (typically data scattered <0.1 log units around the isotherms). However, because (i) often only three measurements were made, (ii) in a limited concentration range, and (iii), isotherms were based on estimated concentrations, the observed nonlinearity should not be extrapolated to concentrations far outside the experimental range. Still, these simple Freundlich isotherms do allow for a good comparison of the sorption affinity between the different mycotoxins based on similar sorbed concentrations in peat.

For several compounds replicate retention data were obtained on a single column with six months in between. Sorption coefficients derived from these two data sets differed by <0.1 log units (see example for aflatoxin B_1 and T-2 toxin in **Figure S4.4**). Retention experiments on columns filled with different peat dilutions with SiC resulted in oc-normalized sorption coefficients that differed

<0.1 log units (e.g. aflatoxin B_1 and resorcylic acid lactones in **Figure S4.4**). The advantage of the experimental design used in this study was, that all compounds were tested using a standard NOM, by a single method that has been demonstrated to be reproducible, accurate, and, as shown in ref[20], relevant for K_{oc} values in most types of soils. Overall, the error margins for the K_{oc} values in this data set (<0.2 log units, see **Table S4.4**) are low compared to the generally observed variety in experimental soil sorption coefficients.[20] Extrapolation of these K_{oc} values to natural soils still needs to be validated for these mycotoxins because other soil constituents may also contribute to the overall soil sorption coefficient.

4.3.2 Measured NOM-sorption affinity of natural toxins

The measured sorption affinity of the neutral mycotoxins varied between log K_{oc} of <0.7 to 3.5, indicating that for only some mycotoxins the sorbed fraction dominates in soils. Although most trichothecenes did not show sufficient retention to quantify the sorption affinity to peat (log K_{oc} <0.7), the only two trichothecene compounds that have the bulky fragment [$OC(=O)CH_2CH(CH_3)_2$] as side chain, HT-2 toxin and T-2 toxin, showed moderate log K_{oc} (**Figure 4.1**, **Table 4.1**), whereas the overall bulky compound verrucarin A had a relatively strong NOM-sorption affinity. The aflatoxins displayed minor differences in sorption affinity for peat; ~0.45 log units higher due to the presence of a double bond both in B_1 compared to B_2, and in G_1 compared to G_2, and 0.3 log units lower due to the presence of an additional OH group in M_1 compared to B_1. The tested resorcylic acid lactones and alternaria toxins have pK_a values well above 7 (**Table 4.1**), and were therefore present in their neutral form in the aqueous eluents of pH 4.5. Of the tested alternaria toxins, altenuene had the strongest sorption affinity to peat. For the resorcylic acid lactones, zearalenone showed slightly stronger sorption compared to the stereoisomers α- and β-zearalenol (**Figure S4.5**). The sorption affinity of zearalenone obtained in this study by using the HPLC-based flow-through method, agreed well to the K_{oc} value generated from batch sorption tests with a natural soil[14] (log K_{oc} values of 3.4, and

3.5, respectively), as well as to feed additive binding studies with leonardite humic acid (derived log K_{oc} average ~3.3[16] and ~3.6[17]). These feed additive binding studies further observed only marginal effects of pH for zearalenone between pH 3-8. The type B trichothecene deoxynivalenol, which did not retain on our columns, also did not show significant binding to humic acids in a feed additive study[16] (derived log K_{oc}<1.6). Aflatoxin B_1 showed slightly stronger binding to the feed additive oxihumate[15] (derived log K_{oc} ~4.3) compared to what we observed for peat (log K_{oc} of 3.5). Zearalenone is currently the only mycotoxin for which measured NOM sorption coefficients are available from various sources, so it remains to be tested how column derived K_{oc} values for other mycotoxins align with K_{oc} values for natural soils. Other soil constituents than NOM may also sorb mycotoxins, since charcoal strongly binds deoxynivalenol,[16] and smectite minerals can also bind zearalenone.[16]

Neutral mycotoxins interact with NOM via van der Waals interactions, H-bonding and other polar interactions, and the pp-LFER concept applies five physico-chemical parameters that can provide clear insight to what extent each interaction contributes to a partitioning coefficient between two phases.[21,37,41,42] Because measured pp-LFER parameters[43] are absent, we tested whether parameters estimated by ABSOLV software[44] (**Table S4.3**) can link NOM sorption coefficients to functional properties of the mycotoxins. Multiple linear regression of measured log K_{oc} values (n=20) with the five ABSOLV parameters (volume, hexadecane-water partitioning coefficient, polarisability, H-bond donor/acceptor capability) results in a poor fit, with R^2 of 0.2 and a RMSE of 0.91 log units. Apparently, these fairly well defined (though calculated from molecular structure) pp-LFER properties do not explain the variance in our measured sorption coefficients. The section on pp-LFER predicted K_{oc} values applies the estimated ABSOLV descriptors in an existing pp-LFER equation for sorption affinity to Pahokee peat.[22]

Table 4.1. Chemical-physical parameters and experimental compound-dependent conditions, including pK_a, the speciation at pH 4.5, and the used detection wavelength λ_{max}, tested aqueous concentration [mg/L], and experimental log K_{oc} data normalized to 100 mg/kg$_{oc}$.

CAS #	COMPOUND	pK_a [a]	speciation at pH 4.5	λ_{max} [nm]	tested range $C_{AQ\text{-}est}$ [mg/L]	exp. log K_{OC} [L/kg$_{oc}$] (*N*) [b]
	Ergot alkaloids					
564-36-3	Ergocornine	7.8	100% cationic	220	0.01 - 1.2	3.88 (5)
511-09-1	Ergocryptine	7.8	100% cationic	205	0.02 - 0.3	4.02 (4)
	Penicillium					
518-75-2	Citrinin	3.1	96% anionic	214	0.2 - 4.4	3.10 (3)
149-29-1	Patulin	12.1	neutral	276	0.1 - 1.5	1.08 (3)
	Aflatoxins					
6795-23-9	Aflatoxin M_1	-	neutral	226	0.7 - 6.2	3.18 (3)
1162-65-8	Aflatoxin B_1	-	neutral	223	0.8 - 10.7	3.46 (12)
7220-81-7	Aflatoxin B_2	-	neutral	216	0.4 - 3.3	3.05 (3)
1165-39-5	Aflatoxin G_1	-	neutral	217	0.7 - 4.3	3.31 (3)
7241-98-7	Aflatoxin G_2	-	neutral	215	0.2 - 3.6	2.80 (3)
	Alternaria toxins					
23452-05-3	Alternariol MME [c]	7.0	99.7% neutral	205	0.01-0.1	1.00 (3)
641-38-3	Alternariol	7.2	99.8% neutral	205	0.1 - 0.6	2.10 (6)
29752-43-0	Altenuene	7.4	99.9% neutral	241	0.1 - 1.3	2.58 (12)
28540-82-1	Tentoxin	8.8	neutral	217	0.1 - 2.9	1.39 (3)
	Resorcylic acid lactones					
17924-92-4	Zearalenone	7.6	99.9% neutral	235	0.2 - 3.2	3.42 (8)
36455-72-8	α-Zearalenol	7.6	99.9% neutral	211	0.1 - 1.2	2.87 (10)
71030-11-0	β-Zearalenol	7.6	99.9% neutral	217	0.4 - 2.1	2.81 (13)
	Trichothecenes A					
21259-20-1	T-2 Toxin	13.2	neutral	205	8.7 - 35.5	1.03 (6)
26934-87-2	HT-2 Toxin	13.3	neutral	205	3.9 - 93.3	1.01 (5)
36519-25-2	Neosolaniol	13.1	neutral	205	3.0 - 8.7	<0.7
2270-40-8	Diacetoxyscirpenol	13.4	neutral	205	1.3 - 22.4	<0.7
2198-92-7	Verrucarol	15.0	neutral	205	0.6 - 4.9	<0.7
02.09.3148	Verrucarin A	14.6	neutral	261	0.1 - 5.0	2.15 (8)
	Trichothecenes B					
50722-38-8	3-Acetyl-DON [d]	11.8	neutral	218	2.7 - 46.8	<0.7
88337-96-6	15-Acetyl-DON [d]	11.8	neutral	223	1.7 - 5.6	<0.7
51481-10-8	DON [d]	11.9	neutral	218	2.0 - 8.5	<0.7
23255-69-8	Fusarenone-X	11.7	neutral	218	5.0 - 51.3	<0.7
23282-20-4	Nivalenol	11.8	neutral	211	7.2 - 20.0	<0.7
	Other mycotoxins					
519-57-3	Sulochrin	7.3	99.8% neutral	207	0.2 - 3.2	1.93 (3)
10048-13-2	Sterigmatocystin	6.9	99.6% neutral	205	0.5 - 1.9	2.07 (3)
	Phytoestrogens					
486-66-8	Daidzein	7.2	99.8% neutral	248	0.1 - 2.2	3.29 (3)
531-95-3	Equol	10.0	neutral	260	0.2 - 3.4	2.78 (3)

[a] pK_a values predicted by SPARC; [b] sorption coefficients are normalized to 100mg/kg$_{OC}$ from fitted exponential sorption isotherms. Uncertainty margins of the experimental log K_{OC} values were determined separately and assessed to be less than 0.2 log units (Table S4), *N*=injected samples; [c] MME = monomethylether; [d] DON = deoxynivalenol.

4.3.3 Measured NOM-sorption affinity of ionized mycotoxins

The negatively charged *Penicillium* compound citrinin (~96% ionic at pH 4.5) sorbed stronger to peat than the neutral *Penicillium* compound patulin (**SI Figure S4.5-B**). It is not clear how much the small neutral fraction of citrinin at pH 4.5 contributes to the overall sorption, but organic anions typically bind 1-2 orders of magnitude weaker than their neutral species,[22] and no effect of pH has been observed for NOM sorption of organic anions in the range of pH 4.5-7.[22] The ergot alkaloids ergocornine and ergocryptine were positively charged (pK_a 8.8), and show the strongest sorption to Pahokee peat of all tested mycotoxins. The strong affinity to sorb to peat is most likely due to a combination of ionic interactions with negatively charged groups in the peat, and nonionic interactions of the rest of the alkaloid structure with the nonionic organic matter structure adjacent to the cation-exchange site, as has been discussed for cationic surfactants[45] and clarithromycin.[25] Ionic strength and the type of dissolved salts can influence the sorption of organic cations to NOM (e.g. factor 10 stronger sorption at 10 times lower Na^+ concentrations for clarithromycin[25]). The effect of pH for cations is probably small, as long as the pH is not too low (i.e. competing H^+ not too high).[46] Organic matter may not be the sole sorbent phase for organic cations in soil, as clay minerals also bear negatively charged surfaces and can be highly abundant. The sorption behavior of ionizable compounds is poorly understood, rendering experimentally derived sorption values even more important.

4.3.4 K_{oc} predictions based on K_{ow} as single descriptive parameter

K_{ow} is a commonly available parameter for neutral contaminants, and forms the central physicochemical property in risk assessment guidelines because for a wide variety of compounds a sufficiently strong correlation was demonstrated between K_{ow} and K_{oc}.[13] Class-specific K_{ow}-K_{oc} relationships can improve the predictive values for compounds with different functional groups,[31] or additional correction factors can be applied (EPISuite's KocWIN). However, since experimental K_{ow} values are absent for all tested toxins, except the phytoestrogens equol and daidzein,[27] K_{ow} needs to be estimated first before it can be applied in a K_{ow}-K_{oc} regression model. We are aware that this is stretching the limits of proper QSAR use. However, since octanol-water is a much simpler, and better calibrated, model system than NOM-water, it was interesting to see how well the outcomes of different predictive K_{ow} models align for mycotoxins, before focusing on predictive K_{oc} models. Octanol is not likely to be a suitable surrogate for the interactions of ionized organics with NOM, because octanol has no way of interacting with ionized organics by coulombic forces, whereas NOM does. In fact, none of the predictive K_{ow} tools, and so far also no predictive K_{oc} tools, has been designed to deal with ionized organics. The remainder of this discussion on predictive tools only involves mycotoxins that are largely neutral at the tested pH range.

Several commonly used, but distinctly different, approaches were applied to calculate K_{ow} for this set of mycotoxins (**Table S4.2**, **Figure S4.6**). For the resorcylic acid lactones and the phytoestrogen equol, the K_{ow} range of all predictions was fairly narrow, with standard deviation of 0.4 log units. The differences between experimental K_{ow} for the two phytoestrogens were also correctly predicted by nearly all estimation tools. For most of the other mycotoxins, however, the predicted K_{ow} ranged over two orders of magnitude. Some extreme outliers are the EPISuite value for verrucarin A, being four orders of magnitude below the lowest other predicted value (included correction factors are -5.4 log units for the special cyclic ester ring structure in addition to -2.9 for three ester fragments), and the COSMO*therm* value for fusarenone X, being three orders of magnitude above the cluster of other predicted values. In the absence of measured K_{ow} values it is difficult to say which predictive model provides the best result for mycotoxins. The variety in predicted K_{ow} values at least shows that most models lack the appropriate correction factors or that their algorithms have not been sufficiently calibrated to deal with such structures.

ALOGPS K_{ow} values showed a strong correlation with experimental K_{ow} data (**Figure S4.7-A**) for a set of complex neutral compounds published recently.[24] In addition, ALOGPS K_{ow} values were correlated with experimental

K_{oc} data (RMSE of 0.68) for a wide variety of neutral compounds when combining two extensive sorption data sets that have used the same micronized Pahokee peat[24,20] (**Figure S4.7-B**). Still, neither ALOGPS nor any of the other tools align their predicted K_{ow} with our experimental K_{oc} data for mycotoxins (**Figure S4.7-C**). Another way of assessing the appropriateness of K_{ow} to predict K_{oc} would be to determine if sorption differences between compounds within a given class of mycotoxins are resembled by predicted K_{ow} values, but no model stands out. The difference between the K_{oc} for aflatoxins is quite well reflected by the K_{ow} from *COSMOtherm*, and not well by EPISuite. Still, opposing trends were observed for the resorcylic acid lactones, and none of the models correctly predicted differences between the phytoestrogens (**Figure S4.6**).

4.3.5 K_{oc} predictions based on fragment contribution values

EPISuite's KocWIN predicted K_{oc} values in two ways (**Table S4.5**); one based on K_{ow} values with additional correction factors, the other applying a molecular connectivity index (MCI) based on experimental K_{oc} data. In both cases, EPISuite predicted values do not correlate well with experimental ones, at least 10 compounds differ more than one order of magnitude (**Figure S4.8**). Still, the non-retained mycotoxins on the peat columns are all predicted to have a low K_{oc} by both methods. Interestingly, the strong K_{ow} outlier verrucarin A (**Table S4.2**) yields a K_{oc} value in a more realistic range using the MCI method. Overall, the application of correction factors to derive K_{oc} values for mycotoxins does not seem to work adequately because the models were (probably) not sufficiently calibrated for the complex functional groups in mycotoxins.

4.3.6 K_{oc} predictions based on multiple physic-chemical descriptors

Recently, a pp-LFER was derived specifically for Pahokee-peat/water sorption using a sorption data set for 129 structurally diverse neutral compounds, including pesticides and pharmaceuticals, that was obtained with the same column method (RMSE of 0.4).[21] The only mycotoxin for which all experimental pp-LFER descriptors have become available is zearalenone (see values in **Table S4.3**).[24,47] For this compound, the pp-LFER predicted Log K_{oc} of 3.61 (using the hexadecane-air partitioning coefficient[21,38]) is very close to our experimental Log K_{oc} of 3.42 (**Table 4.1**). Unfortunately, the suitability of the pp-LFER approach for mycotoxins cannot be properly assessed because experimental pp-LFER descriptors for other mycotoxins are not yet available. However, all pp-LFER descriptors can also be estimated by the program ABSOLV,[44] as presented in **Table S4.3** for those mycotoxins that are neutral at pH 4.5. This will, however, greatly increase uncertainty in pp-LFER predicted K_{oc} values, because estimated descriptors have wider error margins than experimental descriptors. Predicted K_{oc} values obtained with estimated pp-LFER descriptors are summarized in **Table S4.5**, and are plotted against experimental K_{oc} values in **Figure S4.9**. Again, there are more mycotoxins for which K_{oc} predictions deviate more than ten-fold than such which are in close range of experimental values. Nevertheless, all weakly sorbing mycotoxins are also predicted to have low K_{oc}.

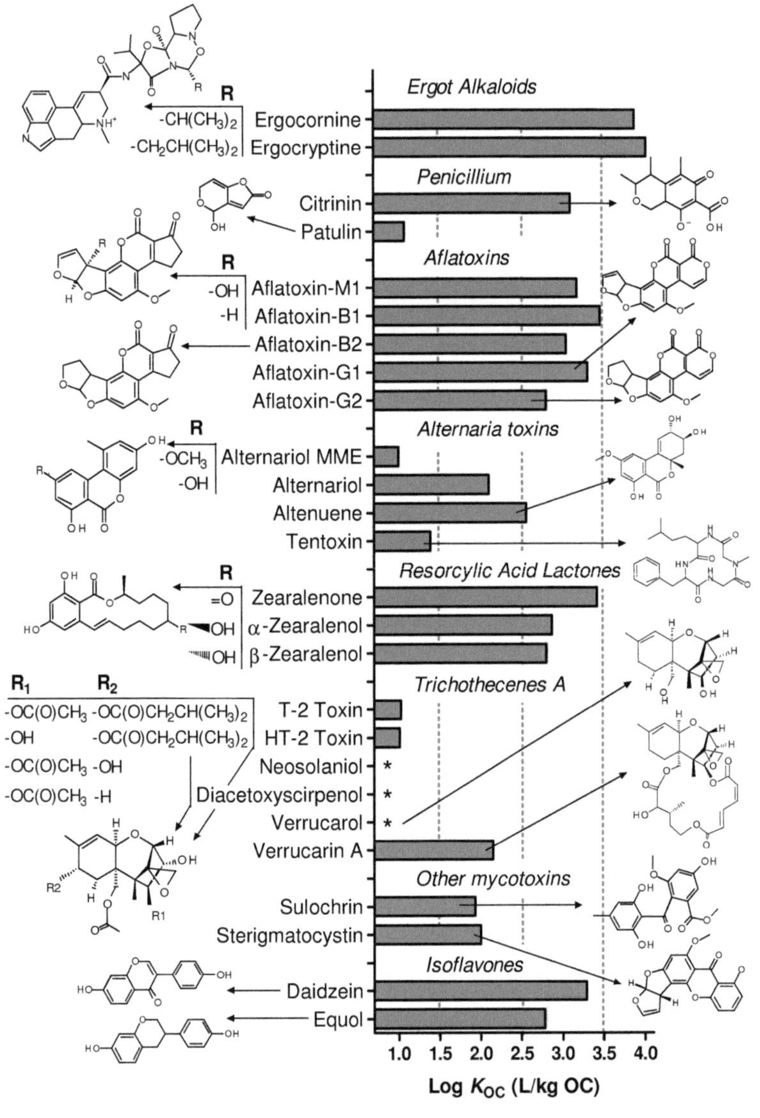

Figure 4.1. Organic carbon normalized sorption coefficients to micronized Pahokee peat for neutral mycotoxins, determined with the column method at pH 4.5, normalized to sorbed concentration of 100 mg/kg$_{oc}$. * Sorption below the lower dynamic range endpoint of log K_{oc} 0.7. All trichothecenes B are below this threshold and therefore not shown. Error margins are not shown, but for all compounds <0.2 log units. Sorption isotherms are in Figure S4.5.

4.3.7 K_{oc} predictions based on NOM models with SPARC

SPARC allows partitioning calculations for any organic phase in SMILES format, as long as it is not too large to render calculation time too long. Therefore, virtually any structure that could represent NOM can be used. Various NOM models and humic acid (HA) model representations have been used recently by Phillips et al.[33] to test the predictive power of both SPARC and *COSMO*-SAC to estimate K_{oc} values for neutral compounds. In the following, both the SPARC and *COSMOtherm* (closely related to *COSMO*-SAC software, discussed further below) predicted sorption coefficients for several NOM models are corrected for oc-content using the molecular weight percentage of the fraction of C atoms. SPARC calculations were limited to the smallest NOM models, and K_{oc} values were therefore calculated for the NOM model representing leonardite (**Table S4.5**), and for two 'terrestrial' HA's and three more 'aquatic' HA's (tested range oc-content 51-63%, **Figure S4.10**).

To assess the influence of NOM structure, for all six NOM models K_{oc} values were calculated for zearalenone, HT-2 and patulin, as well as for a set of eight "simple" nonpolar and polar neutral compounds (**Figure S4.10**). The K_{oc} range calculated by SPARC for the eight simple compounds on the six different NOM models is 0.5-2.1 log units. For the two mycotoxins, this range is comparable for zearalenone (0.7 log units), rather high for the small compound patulin (1.9) and widely scattering for the large multifunctional compound HT-2 toxin (3.6 log units). Not only is the K_{oc} range for mycotoxins among the different NOM models relatively wide, **Figure S4.10** also shows that the different affinities between two NOM models are not necessarily following the same trend, when comparing different compounds. For example, even for simple structures these predictions suggest that between HA-2 and HA-3 the sorption affinity of phenol increases whereas for cyclohexanol the sorption affinity decreases. It is therefore not surprising that finding a NOM model that allows SPARC to adequately predict sorption coefficients of complex molecules is extremely difficult. The K_{oc} values for all neutral (myco)toxins were only calculated for the leonardite NOM model (pH 4.5; **Table S4.5**). The leonardite model is surely not the best fitting model for the experimental mycotoxin data set (**Figure S4.11**). The data in **Figure S4.10** already showed that with the leonardite NOM model the highest K_{oc} values were predicted, and **Figure S4.11** illustrated that for all compounds predicted sorption affinities were far above experimental values. Notably, the non-retaining compounds were predicted amongst the values with strongest affinities. Tuning the NOM model to fit these data better can only be done when the principles underlying the errors are better understood.

4.3.8 K_{oc} predictions based on NOM models with quantum-chemical statistical thermodynamics

K_{oc} predictions by *COSMOtherm* (**Table S4.5**) could be expected to have the same problems as SPARC with finding an optimized NOM model to best represent experimental data. However, despite using the same leonardite NOM model as used for SPARC, *COSMOtherm* predicted K_{oc} values are surprisingly in the same range as experimental values, though the scatter with experimental K_{oc} is comparably high as for K_{oc} predictions with EPISuite or pp-LFER. While aflatoxins and most alternaria toxins were predicted within 0.5 log units, alternariol methylether (MME), the zearalenol isomers and the trichothecenes were predicted to have a much stronger sorption than observed (**Table S4.5** - **Figure S4.11**). Just as for K_{oc} predictions with K_{ow}, also the stronger NOM sorption of daidzein compared to equol was not expected for K_{oc} predictions with *COSMOtherm*. Only half of all non-retaining mycotoxins were also amongst the lowest predicted K_{oc} values, but 12 out of 20 neutral mycotoxins are predicted within a factor 10 of the experimental K_{oc}. Note that the *COSMOtherm* values are not calibrated to a set of known sorption coefficients, but are solely based on statistical-thermodynamics using quantum-chemical properties. Therefore, the offset of the predictions against experimental values is not strictly defined, and *COSMOtherm* predicted data should best be considered on a relative scale. Excluding the non-retaining mycotoxins, a linear fit on data for predicted log K_{oc} vs. experimental log K_{oc} values (fixed slope of 1) resulted in a regression with

RMSE of 1.03 (**Figure S4.11**). Given the narrow window of the experimental log K_{oc} range this would not be adequate for proposing *COSMOtherm* use for risk assessment of complex organic compounds. Still, this RMSE was comparable to the findings by Phillips et al.[33] who applied *COSMOtherm* for several NOM models for a much broader set of K_{oc} values for neutral compounds, including 440 compounds and spanning 6 orders of magnitude. Similar to SPARC, it is not likely that a better representative NOM model, in terms of aligning predicted K_{oc} with experimental K_{oc} values, can easily be derived.

4.3.9 Evaluation of predictive tools and recommendations for future K_{oc} determination/estimation

In the absence of reference partitioning properties, the reported experimental K_{oc} values represent a unique set of sorption affinities for mycotoxins for use in risk assessment. The complex molecular structure of most mycotoxins, and that of NOM structures, is clearly problematic for predictive tools. Even predicted K_{ow} values for mycotoxins vary over two orders of magnitude. None of the predictive K_{ow} tools accounts for the trends observed for K_{oc} amongst the various tested mycotoxins. Direct predictions for K_{oc} are possible as well, but the online tool SPARC and quantum-chemical software approach *COSMOtherm* rely heavily on an appropriate input structure to resemble the NOM in K_{oc} predictions. The pp-LFER approach does not require a NOM input structure because this information is covered by a calibration data set, but it requires experimentally derived sorbate descriptors that are not yet available for mycotoxins. When descriptors become available (or estimated descriptors become validated) for highly complex neutral compounds, the pp-LFER approach may well be the most suitable method to predict partitioning values into complex phases such as NOM. For mycotoxins that are largely ionized under common environmental conditions, no reliable predictive tools exist, and the experimental method presented in this work appears to be a suitable method to deliver precise sorption coefficients within a certain range. Although these NOM sorption values can be regarded as representative K_{oc}

values for most neutral compounds in many natural soils within modest uncertainty margins (~0.6 log units for many neutral compounds[21]), this approach remains to be verified - and can be further optimized - for mycotoxins by measurements on actual soils. Mycotoxin soil sorption affinities can be expected to be higher than measured by the K_{oc} approach, especially when other sorbent phases are present, e.g. in clayey soils with low OM-fractions, or soils rich in black carbon material. Especially for ionized mycotoxins it is not yet clear to what extent peat is a representative NOM and whether other soil constituents play a role in sorption processes. To some extent these uncertainties can be overcome by applying the column method with packings of micronized natural soils, as has been already done for a series of Eurosoils.[21] For some compounds it still remains difficult to obtain K_{oc} values with the column method, because they either sorb too weakly or too strongly (optimized testing range for log K_{oc} lies between ~0.5 - 4.5), or they are retained on the packing material SiC (though other more suitable column packing material can be considered). Optimization with respect to the use of suitable more sensitive analytical detectors, such as FLD or MS/MS, would allow working with lower concentrations, and hence, environmentally (but not necessarily toxicologically) more relevant ranges.

Acknowledgment

We thank the Swiss Federal Office for the Environment for financial support of this study.

References (chapter 4)

1. Kuiper-Goodman, T., *Food safety: Mycotoxins and phycotoxins in perspective*, in *Mycotoxins and Phycotoxins - Developments in Chemistry, Toxicology and Food Safety*, M. Miraglia, et al., Editors. 1998, Alaken, Inc: Ft Collins. p. 25.
2. Hussein, H.S. and Brasel, J.M., Toxicity, metabolism, and impact of mycotoxins on humans and animals. *Toxicology*, **2001**. 167, 101 - 134.
3. Nganje, W.E., et al., Regional economic impacts of Fusarium head blight in wheat and barley. *Rev. Agric. Econ.* **2004**. *26*. 332 - 347.

4. Krska, R., et al. *Mycotoxin analysis: An update*. in *12th IUPAC International Symposium on Mycotoxins and Phycotoxins*. 2007. Istanbul, TURKEY: Taylor & Francis Ltd.
5. Shephard, G.S., Determination of mycotoxins in human foods. *Chemical Society Reviews*, **2008**. *37*. 2468-2477.
6. Bucheli, T.D., et al., *Fusarium* mycotoxins: overlooked aquatic micropollutants? *J. Agric. Food Chem.* **2008,** *56*, 1029-1034.
7. Hartmann, N., Occurrence of zearalenone on *Fusarium graminearum* infected wheat and maize fields in crop organs, soil and drainage water. *Environ. Sci. Technol.* **2008,** *42*, 5455-5460.
8. Hartmann, N., et al., Quantification of zearalenone in various solid agroenvironmental samples using D-6-zearalenone as the internal standard. *J. Agric. Food Chem.* **2008,** *56*, 2926-2932.
9. Gilbert, J., et al. *Assessment of dietary exposure to ochratoxin A in the UK using a duplicate diet approach and analysis of urine and plasma samples*. in *1st Symposium on Risk Assessment and Communication for Food Safety*. 2000. York, England: Taylor & Francis Ltd.
10. Turner, P.C., et al., Urinary deoxynivalenol is correlated with cereal intake in individuals from the United Kingdom. *Environ. Health Persp.* **2008,** *116*, 21 - 25.
11. Wettstein, F.E. and Bucheli, T.D., Poor elimination rates in waste water treatment plants lead to continuous emission of deoxynivalenol into the aquatic environment. *Water Res.* **2010,** *44*, 4137-4142.
12. Schenzel, J., et al., A multi-residue screening method to quantify mycotoxins in aqueous environmental samples. *J. Agric. Food Chem.* **2010,** *58*, 11207-11217.
13. Schwarzenbach, R.P., et al., *Sorption I: Sorption processes involving organic matter*, in *Environmental organic chemistry*. 2003, John Wiley & Sons, Inc.: New York.
14. Hartmann, N. Thesis: Environmental exposure to estrogenic mycotoxins. Swiss Federal Institute of Technology, Zurich, 2008.
15. Jansen Van Rensburg, C., et al., In vitro and in vivo assessment of humic acid as an aflatoxin binder in broiler chickens. *Poultry Science* **2006,** *85*, 1576-1583.

16. Sabater-Vilar, M., et al., In vitro assessment of adsorbents aiming to prevent deoxynivalenol and zearalenone mycotoxicoses. *Mycopathologia* **2007,** *163*, 81-90.
17. Santos, R.R., et al., Isotherm modeling of organic activated bentonite and humic acid polymer used as mycotoxin adsorbents. *Food Additives & Contaminants: Part A* **2011,** *28*, 1578-1589.
18. Bi, E., et al., Sorption of heterocyclic organic compounds to reference soils: Column studies for process identification. *Environ. Sci. Technol.* **2006,** *40*, 5962-5970.
19. Mader, B.T., et al., Sorption of nonionic, hydrophobic organic chemicals to mineral surfaces. *Environ. Sci. Technol.* **1997,** *31*, 1079-1086.
20. Bronner, G. and Goss, K.U., Sorption of organic chemicals to soil organic matter: influence of soil variability and pH dependence. *Environ. Sci. Technol.* **2011,** *45*, 1307-1312.
21. Bronner, G. and Goss, K.U., Predicting sorption of pesticides and other multifunctional organic chemicals to soil organic carbon. *Environ. Sci. Technol.* **2011,** *45*, 1313-1319.
22. Tulp, H.C., et al., pH-Dependent sorption of acidic organic chemicals to soil organic matter. *Environ. Sci. Technol.* **2009,** *43*, 9189-9195.
23. Tolls, J., Sorption of veterinary pharmaceuticals in soils: A review. *Environ. Sci. Technol.* **2001,** *35*, 3397-3406.
24. Tulp, H.C., et al., Experimental determination of LSER parameters for a set of 76 diverse pesticides and pharmaceuticals. *Environ. Sci. Technol.* **2008,** *42*, 2034-2040.
25. Sibley, S.D. and Pedersen, J.A., Interaction of the macrolide antimicrobial clarithromycin with dissolved humic acid. *Environ. Sci. Technol.* **2008,** *42*, 422-428.
26. Karickhoff, S.W., Semiempirical estimation of sorption of hydrophobic pollutants on natural sediments and soils. *Chemosphere* **1981,** *10*, 833-846.
27. Rothwell, J.A., et al., Experimental determination of octanol-water partition coefficients of quercetin and related flavonoids. *J. Agric. Food Chem.* **2005,** *53*, 4355-4360.
28. EPA, U.S. Estimation Program Interface (EPI) Suite, v4.10. (January 25), United States Environmental Protection Agency, Washington, DC, USA

29. Karickhoff, S.W., et al. SPARC On-line Calculator v4.5. (http://archemcalc.com/sparc/ January-February 2011), Athens, GA
30. Endo, S., et al., LFERs for soil organic carbon-water distribution coefficients (Koc) at environmentally relevant sorbate concentrations. *Environ. Sci. Technol.* **2009,** *43*, 3094-3100.
31. Nguyen, T.H., et al., Polyparameter Linear Free Energy Relationships for estimating the equilibrium partition of organic compounds between water and the natural organic matter in soils and sediments. *Environ. Sci. Technol.* **2005,** *39*, 913-924.
32. Niederer, C. and Goss, K.U., Quantum chemical modeling of humic acid/air equilibrium partitioning of organic vapors. *Environ. Sci. Technol.* **2007,** *41*, 3646-3652.
33. Phillips, K.L., et al., Prediction of soil sorption coefficients using model molecular structures for organic matter and the quantum mechanical *COSMO*-SAC model. *Environ. Sci. Technol.* **2011,** *45*, 1021-1027.
34. Wittekindt, C. and Goss, K.U., Screening the partition behavior of a large number of chemicals with a quantum-chemical software. *Chemosphere* **2009,** *76*, 460-464.
35. Bradford, G.R., et al., Trace and major element contents of soil saturation extracts. *Soil Sci.* **1971,** *112*, 225-230.
36. Valocchi, A.J., Validity of the local equilibrium assumption for modeling sorbing solute transport through homogeneous soils. *Water Resources Res.* **1985,** *21*, 808-820.
37. Abraham, M.H., et al., Hydrogen-Bonding 34. The factors that influence the solubility of gases and vapors in water at 298-K, and a new method for its determination. *Journal of the Chemical Society, Perkin Transactions 2* **1994**, 1777-1791.
38. Bronner, G., Thesis. Sorption of polar and nonpolar organic compounds in soil organic matter: pH-dependence, sorbent variability and extension of pp-LFER models to multifunctional compounds. Swiss Federal Institute of Technology Zurich, Zurich, 2010.
39. Doucette, W.J., Quantitative structure-activity relationships for predicting soil-sediment sorption coefficients for organic chemicals. *Environ. Toxicol. Chem.* **2003,** *22*, 1771-1788.

40. Goss, K.U., et al., Nonadditive effects in the partitioning behavior of various aliphatic and aromatic molecules. *Environ. Toxicol. Chem.* **2009,** *28*, 52-60.

41. Goss, K.U. and Schwarzenbach, R.P., Linear free energy relationships used to evaluate equilibrium partitioning of organic compounds. *Environ. Sci. Technol.* **2001,** *35*, 1-9.

42. Goss, K.-U., Predicting the equilibrium partitioning of organic compounds using just one linear solvation energy relationship (LSER). *Fluid Phase Equil.* **2005,** *233*, 19-22.

43. Abraham, M.H., et al., Determination of sets of solute descriptors from chromatographic measurements. *J. Chromatogr. A* **2004,** *1037*, 29-47.

44. ABSOLV ACD/ADME Suite - Advanced Chemistry Development. Toronto, Ontario, Canada

45. Ishiguro, M., et al., Binding of cationic surfactants to humic substances. *Colloid Surf. A* **2007,** *306*, 29-39.

46. Brownawell, B.J., et al., Adsorption of organic cations to natural materials. *Environ. Sci. Technol.* **1990,** *24*, 1234 - 1241.

47. Bronner, G., et al., Hexadecane/air partitioning coefficients of multifunctional compounds: Experimental data and modeling. *Fluid Phase Equil.* **2010,** *299*, 207-215.

Supporting Information

Section SI-4.1 Chemicals and properties of Pahokee peat

Chemicals included in this study: 3-acetyl-deoxynivalenol,15-acetyl-deoxynivalenol, aflatoxin B_1 (≥99%), aflatoxin B_2 (≥98%), aflatoxin G_1 (≥99%), aflatoxin G_2 (≥99%), aflatoxin M_1 (≥99%), citrinin (≥99%), deoxynivalenol (≥99%), HT-2 toxin (≥99%), patulin (≥99%) and T-2 toxin (≥99%) which were obtained from Fermentek (Jerusalem, Israel). Alternariol (≥96%), alternariol monomethylether (purity [n.a.]), altenuene (purity [n.a.]), beauvericin (≥97%), diacetoxyscirpenol (≥99%), ergocornine (≥95%), fusarenon-X (purity [n.a.]), neosolaniol (purity [n.a.]), nivalenol (purity [n.a.]), sterigmatocystin (≥98%), sulochrin (≥98%), tentoxin (≥99%), verrucarin A (≥98%), zearalenone (≥99%), α-zearalenol (≥98%), and β-zearalenol (≥95%) were supplied by Sigma (Sigma-Aldrich GmbH, Buchs, Switzerland). Ergocryptine (purity [n.a.]) was purchased from Biopure (Referenzsubstanzen GmbH, Tulln, Austria). Two selected isoflavones, daidzein (≥98%) and equol (≥99%), were supplied by Fluka AG (Buchs, Switzerland).

Table S4.1: Properties of Pahokee peat bulk material including the elemental composition

Soil organic matter	Identity IHSS	Rel. amount in bulk material	Rel. amount in dry material	Relative amount in ash free sample % (w/w)					
		H_2O % (w/w)	Ash % (w/w)	C	H	O	N	S	P
Pahokee Peat II bulk material	PPII 2S 103P	6.2	12.7	46.9	3.9	30.3	3.4	0.6	n.d.

Pahokee Peat was obtained from the International Humic Substance Society (IHSS) and the elemental analyses was performed by Huffman Laboratories, Wheat Ridge, CO, USA; Isotopic analyses by Soil Biochemistry Laboratory, Dept. of Soil, Water, and Climate, University of Minnesota, St. Paul, MN, USA. The H_2O content is the % (w/w) of H_2O in the air-equilibrated sample (as a function of relative humidity). Ash is the % (w/w) of inorganic residue in a dry sample. C, H, O, N, S, and P are the elemental composition in % (w/w) of a dry, ash-free sample. The data for bulk source materials are reported and %H and

%O are corrected for water content. n.d.: not determined. Organic carbon normalized sorption coefficients are corrected for ash content of the raw organic matter; similar properties are assumed for the micronized batches. Micronization was done by Bronner and Goss[1] applying a peat suspension repeatedly through a smaller opening in a high pressure homogenization device, after which the remaining suspension way freeze-dried, resulting in a powder of particles smaller than 20 μm.

Figure S4.1: Molecular structures of the neutral form of tested mycotoxins

Structures of ergot alkaloids (both largely positively charged at pH4.5):

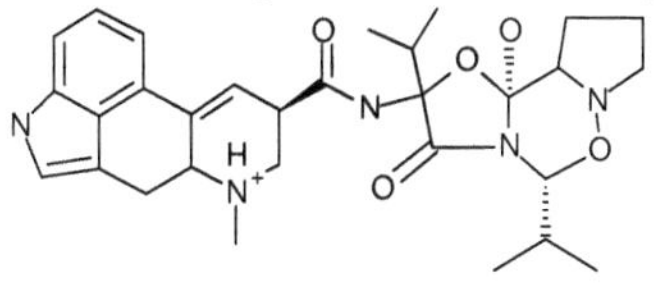

Ergocornine

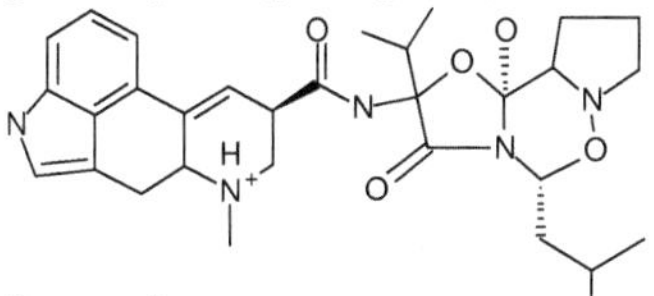

Ergocryptine

Structures of Penicillium toxins (citrinin largely negatively charged at pH 4.5, patulin neutral):

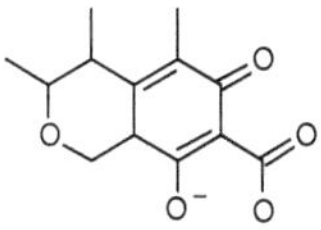

Citrinin

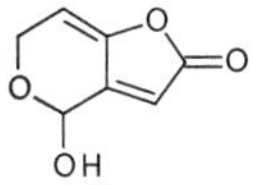

Patulin

Structures of aflatoxins (all neutral at pH 4.5):

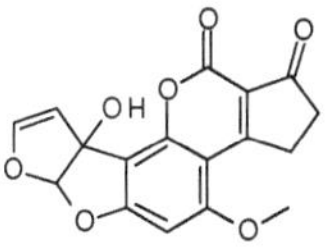

Aflatoxin M_1

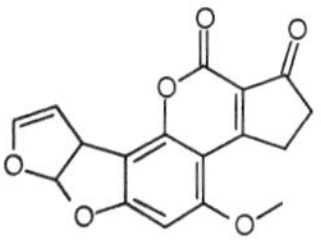

Aflatoxin B_1

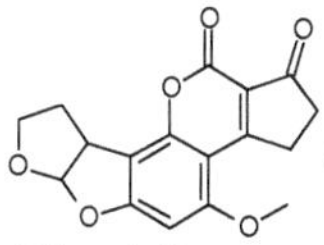

Aflatoxin B_2

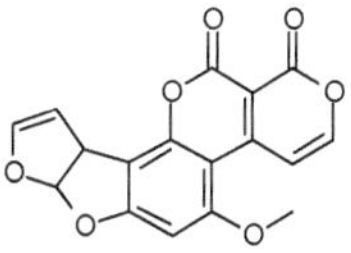

Aflatoxin G_1

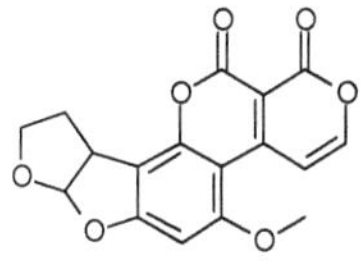

Aflatoxin G_2

Structures of Alternaria toxins (all largely neutral at pH4.5):

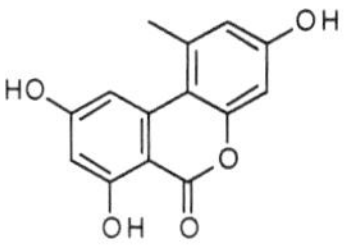

Alternariol

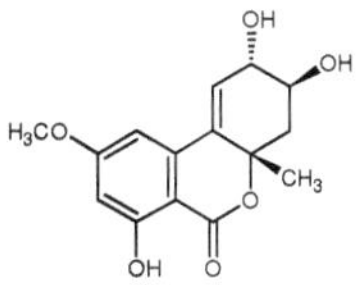

Alternariol monomethylether

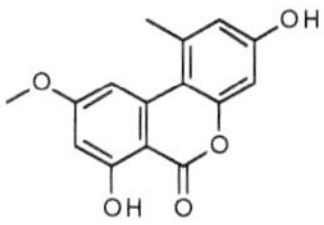

Altenuene

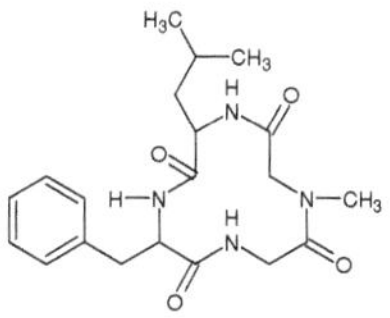

Tentoxin

Structures of resorcylic acid lactones (all largely neutral at pH 4.5):

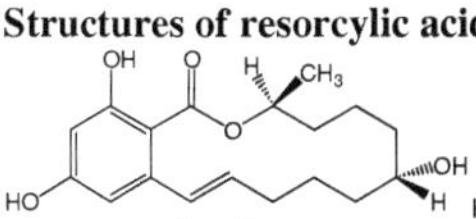

α-zearalenol

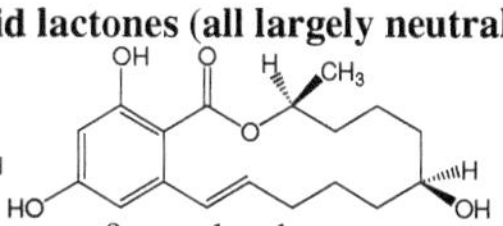

β-zearalenol

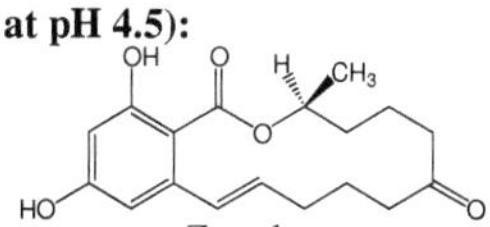

Zearalenone

Structures of trichothecenes (all neutral at pH 4.5):

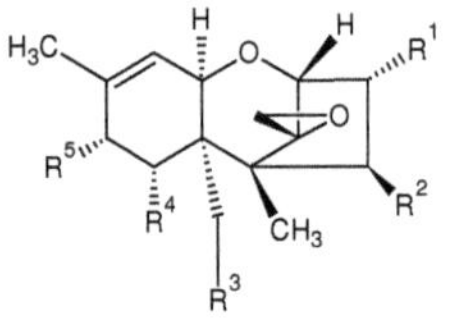

Trichothecenes	R^1	R^2	R^3	R^4	R^5
Type A					
T-2 toxin	OH	$OCOCH_3$	$OCOCH_3$	H	$OCOCH_2CH(CH_3)_2$
HT-2 toxin	OH	OH	$OCOCH_3$	H	$OCOCH_2CH(CH_3)_2$
Diacetoxyscirpenol	OH	$OCOCH_3$	$OCOCH_3$	H	H
Neosolaniol	OH	$OCOCH_3$	$OCOCH_3$	H	OH
Type B					
3-Acetyldeoxynivalenol	$OCOCH_3$	H	OH	OH	=O
15-Acetyldeoxynivalenol	OH	H	$OCOCH_3$	OH	=O
Deoxynivalenol	OH	H	OH	OH	=O
Fusarenon-X	OH	$OCOCH_3$	OH	OH	=O
	OH	OH	OH	OH	=O

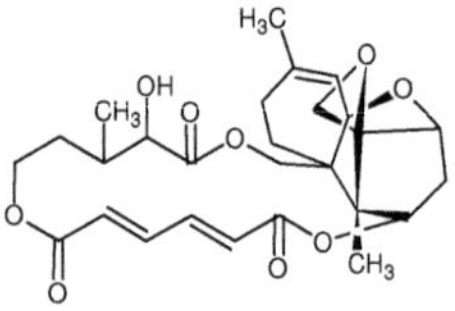

Verrucarin A

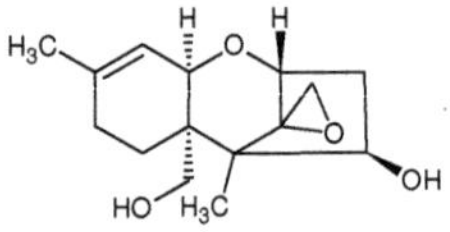

Verrucarol

Structures of "other" mycotoxins (all largely neutral at pH4.5):

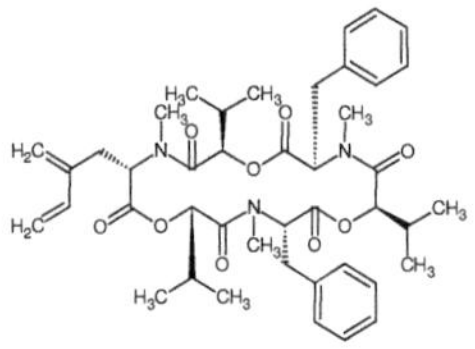

Beauvericin

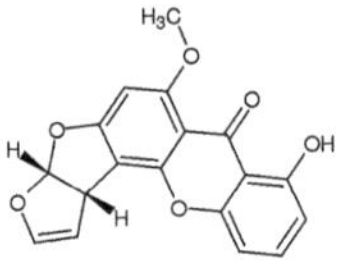

Sterigmatocystin

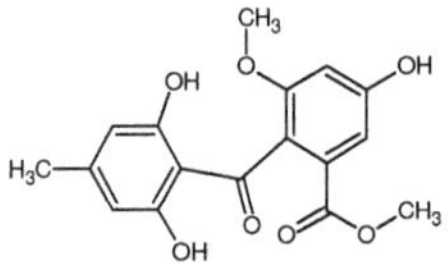

Sulochrin

Structures of selected phytoestrogens (both largely neutral at pH4.5):

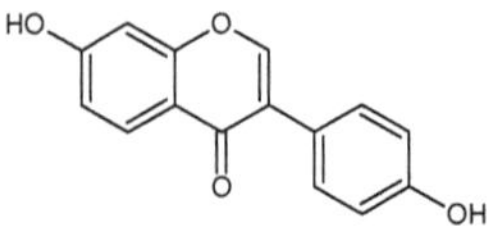

Daidzein

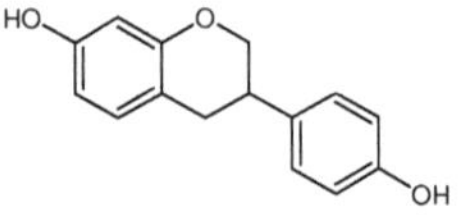

Equol

Table S4.2. Log K_{OW} predictions for mycotoxins tested in this study, which are neutral at pH 4.5

Compound	**ALOGPS**	**ACDLabs**	**pp-LFER via** ABSOLV	**EPIsuite 4.0**	**MARVIN**	**MARVIN**	**MARVIN**	**SPARC**	***COSMO therm***
	v2	**Free 12**	*******	**SRC Physprop data base**	**Phys**	**KloP**	**VG**	**v4.5**	**V2**
Neutral mycotoxins pH 4.5									
Penicillium									
Patulin	-0.72	-0.75	-0.43	-2.4	-0.1	-0.41	-0.62	-1.82	-0.13
Aflatoxins									
Aflatoxin M_1	0.57	-0.35	0.04	-0.27	1.11	1.39	0.29	2.83	0.87
Aflatoxin B_1	1.29	0.45	0.96	1.23	1.5	2.17	1	2.93	0.92
Aflatoxin B_2	1.33	0.37	2.06	1.45	1.84	1.78	1.09	3.31	0.54
Aflatoxin G_1	1.14	-0.17	0.14	0.5	1.44	1.47	1.19	2.93	0.43
Aflatoxin G_2	1.16	-0.25	0.55	0.71	1.61	1.4	1.06	2.45	0.04
Alternaria toxins									
Alternariol MME*	2.74	3.68	3.71	2.91	3.66	3.65	2.66	3.07	3.42
Alternariol	2.4	2.93	2.95	2.35	3.47	3.44	2.63	2.29	2.77
Altenuene	0.79	1.95	2.63	2.23	1.37	0.99	0.66	0.31	1.74
Tentoxin	1.23	1.24	0.13	1.46	1.05	0.25	0.34	-0.53	1.12
Resorcylic acid lactones									
Zearalenone	3.39	3.83	4.5	3.58	4.28	4.47	4.38	3.32	3.63
α-Zearalenol	3.66	4.17	5.25	4.09	4.39	4.24	3.87	3.26	4.03
β-Zearalenol	3.66	4.17	5.25	4.09	4.39	4.24	3.87	3.26	3.9
Trichothecenes A									
T-2 Toxin	1.83	2.25	4.13	2.27	1.25	1.26	0.55	3.49	2.81
HT-2 Toxin	1.27	2.27	3.65	1.67	0.87	0.75	0.12	2.49	2.15
Neosolaniol	-0.07	-0.07	1.96	0.27	-0.55	-0.77	-1.24	0.93	1.36
Diacetoxyscirpenol	0.99	1.33	2.92	1.4	0.62	0.54	-0.03	2.7	1.8

Compound	ALOGPS v2	ACDLabs Free 12	pp-LFER via ABSOLV ***	EPIsuite 4.0 SRC Physprop data base	MARVIN Phys	MARVIN KloP	MARVIN VG	SPARC v4.5	*COSMO therm* V2
Verrucarol	0.61	0.7	2.88	0.48	0.52	0.19	-0.16	1.24	1.61
Verrucarin A	1.35	0.45	5.22	-3.32	2.42	2.31	2.94	3.65	3
Trichothecenes B									
3-Acetyl-DON	-0.48	-0.76	1.39	-0.55	-0.54	-0.4	-0.65	0.24	1.31
15-Acetyl-DON	-0.48	-0.63	1.13	0.3	-0.54	-0.4	-0.65	0.29	-0.21
Deoxynivalenol (DON)	-0.93	-1.41	0.42	-0.71	-0.92	-0.91	-1.07	-1.18	0.09
Fusarenone-X	-1.24	-1.05	0.03	-1.24	-1.41	-1.36	-1.57	-1.21	1.44
Nivalenol	-1.66	-0.75	-0.44	-2.24	-1.79	-1.88	-2	-2.61	-0.02
Other mycotoxins									
Sulochrin	2.87	1.62	2.87	3.81	3.19	3.49	2.62	4.3	3.45
	2.8	3.84	3.22	4.37	4.16	4.31	4.08	2.45	3.23
Sterigmatocystin									
Phytoestrogens **									
Daidzein (exp. K_{OW} = 2.5)	2.5	0.45	1.39	2.55	2.49	2.61	3.09	2.88	1.95
Equol (exp. K_{OW} = 3.2)	3.09	2.98	3.07	3.67	3.16	3.19	3.22	3.87	2.86

* MME monomethylether; ** experimental K_{OW} values measured by flask method[2], *** based on ABSOLV estimated parameters using pp-LFER for K_{ow} from [3]

Table S4.3. Predicted substance descriptors for neutral mycotoxins (at pH 4.5) for pp-LFER equation estimated using ABSOLV software.

COMPOUND	E	S	A	B	V	L
Penicillium						
Citrinin	1.66	2.06	0.63	1.71	1.80	9.31
Patulin	1.00	1.26	0.31	1.00	0.98	5.23
Aflatoxins						
Aflatoxin M_1	2.41	2.24	0.21	1.98	2.07	11.98
Aflatoxin B_1	2.28	2.27	0	1.62	2.01	11.67
Aflatoxin B_2	2.01	2.06	0	1.38	2.06	11.23
Aflatoxin G_1	2.19	3.12	0	1.65	2.07	11.73
Aflatoxin G_2	2.04	3.03	0	1.58	2.11	11.74
Alternaria toxins						
Alternariol MME	2.15	1.77	0.63	0.84	1.89	10.33
Alternariol	2.31	1.89	1.30	0.90	1.75	10.31
Altenuene	1.82	1.76	0.67	1.25	2.03	10.45
Tentoxin	2.21	4.27	0.51	2.63	3.28	17.25
Resorcylic acid lactones						
Zearalenone – ABSOLV	1.25	1.67	0.84	1.12	2.46	11.79
Zearalenone – ***Exp. values*** *	1.25	***1.68***	***0.87***	***1.24***	2.46	***11.95***
α-Zearalenol	1.61	1.67	1.11	1.02	2.51	11.81
β-Zearalenol	1.61	1.67	1.11	1.02	2.51	11.81
Trichothecenes A						
T-2 Toxin	1.49	2.04	0.19	2.18	3.41	14.52
HT-2 Toxin	1.59	1.97	0.50	2.03	3.11	13.60
Neosolaniol	1.62	1.91	0.50	2.08	2.69	12.39
Diacetoxyscirpenol	1.41	1.66	0.19	1.78	2.63	11.67
Verrucarol	1.43	1.23	0.63	1.22	1.98	9.27
Verrucarin A	1.86	2.23	0.17	2.12	3.64	15.62
Trichothecenes B						
3-Acetyl-DON	1.63	1.89	0.48	1.88	2.35	10.84
15-Acetyl-DON	1.67	1.83	0.40	1.98	2.35	10.93
Deoxynivalenol (DON)	1.81	1.70	0.71	1.92	2.05	10.17
Fusarenone-X	1.90	2.05	0.67	2.33	2.41	11.70
Nivalenol	2.00	1.98	0.98	2.18	2.11	10.78
Other mycotoxins						
Beauvericin**	3.16	5.44	0	3.41	6.20	28.15
Sterigmatocystin	2.57	2.24	0.13	1.24	2.11	12.31
Sulochrin	2.09	2.32	1.21	1.31	2.35	12.35
Phytoestrogens						
Daidzein	2.21	2.23	1.16	1.27	1.79	10.34
Equol	1.87	1.68	1.00	0.92	1.81	9.66

* descriptors S, A, B are obtained from experimental work by Tulp et al[4] and refined by Bronner and Goss[1], who also determined the L value[1] **Sorption affinity not tested in this study

Section SI-4.2 Preparation of working solutions for the column experiments and detailed procedure for the pH determination of the solution in the column

Individual stock solutions of all compounds were prepared in pure acetonitrile (MeCN), except Fusarenon-X, and Neosolaniol, which were dissolved in a MeCN/Milli-Q water (1/1, v/v) mixture. Stock solution concentrations ranged from 100 to 1000 mg/L, and were all further diluted at least 100 times with 5 mM $CaCl_2$ aqueous eluents solution to achieve lower concentrations to be injected in the HPLC system via a 200 μL stainless steel loop.

One way of assessing that the micronized peat in the HPLC column was equilibrated with the pH of the mobile phase is to determine the pH in the eluents before and after the column. External UV measurement was performed to determine the pH using an indicator substance (trithiocyanuric acid, TCA, pK_a = 5.12), by measuring the UV absorption wavelengths for TCA at 300 nm (peak for the neutral molecule), and 330 nm (peak for the anion) as described in ref.[5,6]

Section SI-4.3 Detailed description of the sorption experiment, the determination of the sorption coefficient and the quality assurance

First, the system hold-up volume (i.e., void volume of the whole chromatographic setup, including tubing and column void volume) was determined by injecting a non-sorbing tracer compound, $NaNO_3$ with a concentration of 0.5 mg/L, on the SiC column ($V_{SiC\ tracer}$). In addition, the V_{net} retention volume of any given compound *i*, was determined on the SiC column as well ($V_{SiC\ i}$). Using equation SI-1 the system hold-up volume of compound *i* is given by:

$$V_{ret\ i\ SiC} = V_{SiC\ i} - V_{SiC\ tracer} \qquad \text{(SI-1)}$$

Subsequently, the retention volume of the tracer ($NaNO_3$, $V_{Peat\ tracer}$) determined on the Peat column was quantified. Additionally, the retention volume of compound *i* on the Pahokee Peat column was measured. The retention volume of compound *i* on the Pahokee Peat column is then determined with equation SI-2:

$$V_{ret\ i\ Peat} = V_{Peat\ i} - V_{Peat\ tracer} \qquad \text{(SI-2)}$$

The final retention volume of a compound *i*, derived from the equations above, is therefore given by equation SI-3:

$$V_{net\ i\ final} = V_{ret\ i\ Peat} - V_{ret\ i\ SiC} \qquad \text{(SI-3)}$$

Quality assurance: The conservation of the amount of injected chemical mass was determined by recovery studies. Recoveries were calculated from substance specific UV-chromatogram signals of the injected sorbate and the followed integration of the peak. The obtained peak area ($AREA_{peak}$) was compared with a theoretical value ($AREA_{theo}$).

$$AREA_{peak} = Q * \int ABS(t) * dt$$

$$AREA_{theo} = V_{loop} * ABS(c_0)$$

Q is the flow rate used during the recovery study; V_{loop} is the injected volume of the sorbate with the concentration c_o. ABS (c_0) is the UV-absorbance at the concentration c_o, which was externally determined under the use of the detection wavelength. ABS (t) is the UV-absorbance at the same wavelength as applied in the column experiments. All obtained recovery values were in the range of 93-102%.

Overall, sorption to the chromatographic system (capillaries, frits, column wall) and SiC should be, and was, always considerably smaller compared to sorption to the packed amounts of peat. Still, to account for this, retention volume on a 1cm column filled only with SiC was determined for each compound in comparison with retention volume of the non-retaining tracer $NaNO_3$ ($V_{SiC\ i}$ - $V_{SiC\ tracer}$), and subtracted from retention volume on the test column ($V_{ret\ i\ Peat}$ - $V_{Peat\ tracer}$). Uncertainty on the calculated K_{oc} values increases considerably when measured net retention volumes ($V_{net\ i\ final}$) become very small compared to the non-retaining tracer ($V_{Peat\ tracer}$). Therefore, all data for which $V_{net\ i\ final}$ was below 30% of $V_{Peat\ tracer}$ were discarded. This results in a lower detection limit of Log K_{oc} of 0.7 with the 5 cm 10:100 column.

Creating semi-quantitavive sorption isotherms from retention data.

In order to create sorption isotherms, a representative dissolved concentration must be derived from each eluted peak. Three ways were assessed: (i) the peak maximum of each UV-signal recording was easily defined, and therefore these maximum peak absorption values were converted to dissolved concentrations by a single external calibration on a spectrophotometer. (ii) Similarly, the absorption signal of the calculated first moment, typically situated after the peak maximum at a 2-4 times lower intensity (see **Figure S4.2**), can be used, but depends to some extent on the way the peak is integrated. (iii) Bucheli and Gustafsson[7] presented a way to derive an average dissolved concentration from an eluted peak, by analogy between a column (with an effective plate number probably near one) and a mixed batch reactor (see comparison with peak maximum in **Figure S4.3**). Assuming generic Freundlich isotherms to fit the data, the obtained data for different concentrations should fit to a linear line on a logarithmic isotherm plot. **Figures S4.2**, **S4.4** and **S4.5** show that exponential

Freundlich isotherms indeed fit well to all sorption data for the tested mycotoxins obtained from the HPLC column method.

Using the eluted peak maximum to represent aqueous concentrations in order to create Freundlich isotherms is a strongly simplified approach, but the resulting concentration-normalized K_{oc} values show good agreement (root mean square error (RMSE) of 0.15) with the more advanced approach of creating isotherms based on average dissolved concentrations from eluted peaks as presented by Bucheli and Gustafsson[7] (see **Figure S4.3**, the latter approach requires information on dilution factors between injected concentrations which were only available for 5 compounds).

Sorption coefficients were not derived from isotherm data for which fitted Freundlich exponents were considerably below 0.7 or above 1.1. In such cases, either more data points were obtained to reduce the influence of outliers, or the measured sorption coefficients from different samples were averaged to a single value, and as such considered to be a representative sorption value for the tested concentration range only.

Figure S4.2. **A**) Overlay of chromatograms of 3 injections of different concentrations of Altenuene (detected at 241 nm) on HPLC column with peat (3:100 with SiC), 1 injection of non-retaining tracer $NaNO_3$ (detected at 207 nm) on same 3:100 column, 1 injection of Altenuene on column with only SiC (0:100). Red squares indicate peak first statistical moments. **B**) Same overlay as in Figure S2-A, but with time on X-axis plotted in logarithmic scale. **C**) Same plot as in Figure S2-B, but with data for Verrucarin A (detected at 261 nm). **D**) Sorption isotherm calculated for Altenuene from data in plot S2-B, using either peak maximum as dissolved concentration (red dots) or taken from peak 1st moment (open square). **E**) Same plot as Figure S2-D, but with data for Verrucarin A taken from Figure S2-C.

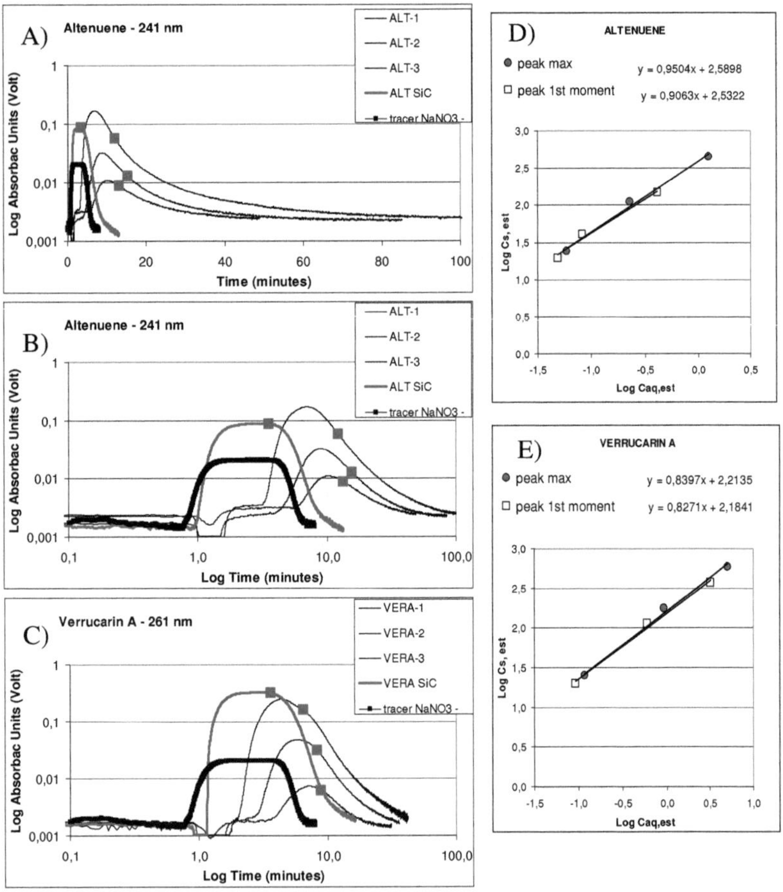

Figure S4.3. Experimentally determined log K_{oc} values (for aflatoxin B_1, altenuene, ergocryptine, zearalenone, verrucarin A) derived from the peak maximum of each individual UV-signal (converted to dissolved concentrations by a single external calibration on a spectrophotometer as in ref [5]) plotted against experimentally determined log K_{oc} values derived from average dissolved concentrations from eluted peak as in ref[7] (by analogy between a column, with an effective plate number probably near one, and a mixed batch reactor). All values given are aligned to a K_{oc} normalized to 100 mg/kg$_{OC}$ using the Freundlich parameters fitted in both ways (R^2 = 0.9331; RSME 0.34).

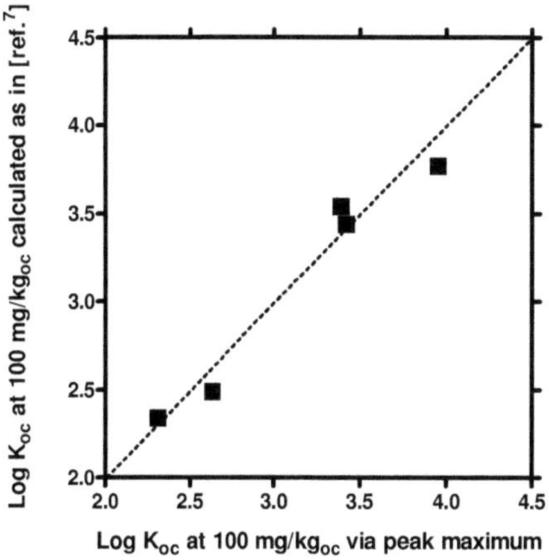

Figure S4.4. Overlap between retention data for two measurement periods (July and December 2010) and columns with different packing densities: **(Left)** Aflatoxin B_1 1:100 and 1.5:100 peat:SiC in 1 cm columns, **(Middle)** for T-2 toxin on a 5 cm long 10:100 column, **(Right)** Resorcylic acid lactones compounds on 1:100 and 3:100 columns.

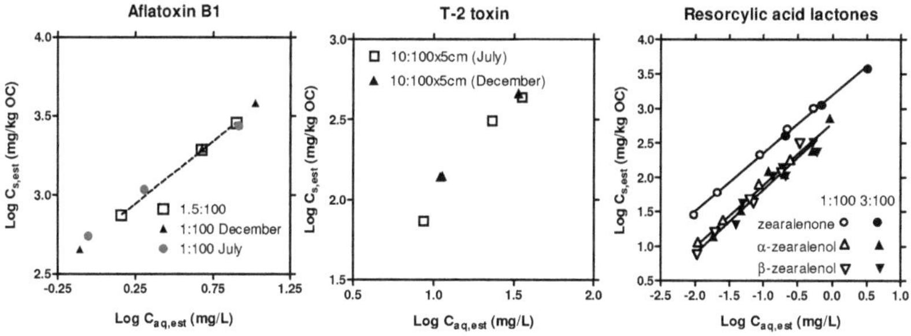

Figure S4.5. Sorption data for different toxins, lines reprent the fitted Freundlich isotherms, dotted lines indicate 1:1 linearity. **A**) Ergot alkaloids, which are both cationic at pH 4.5; **B**) Penicillines, of which citrinin is negatively charged at pH 4.5, while patulin is a neutral mycotoxin. **C**) Aflatoxins; **D**) Alternaria toxins **E**) Resorcylic acid lactones; **F**) Trichothecenes A for which significant retention, in comparison with a nonretaining tracer, was observed on the peat columns; **G**) phytoestrogens.

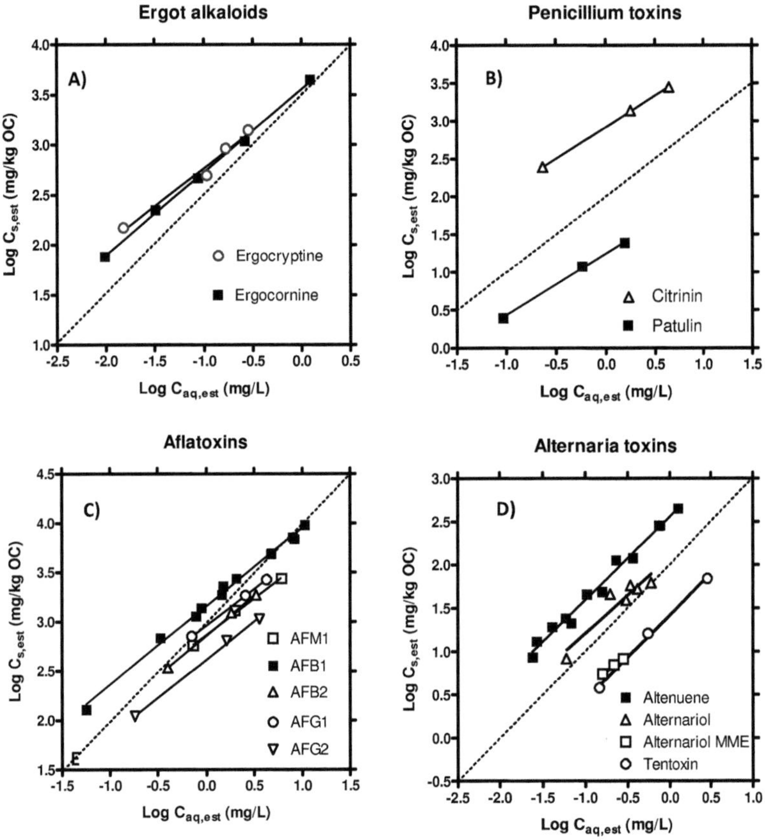

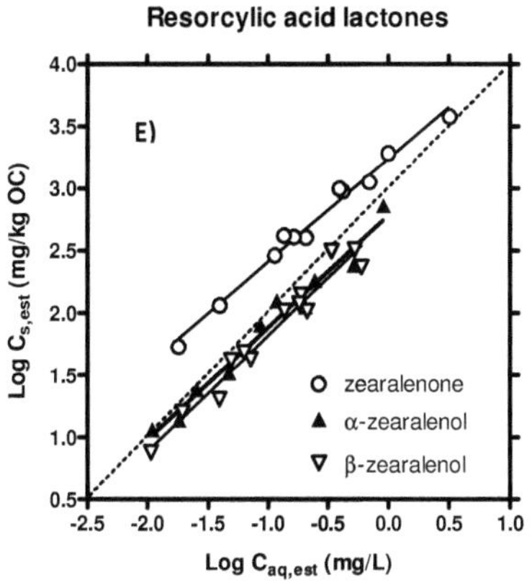
Resorcylic acid lactones
E)
zearalenone
α-zearalenol
β-zearalenol
Log Cs,est (mg/kg OC)
Log Caq,est (mg/L)

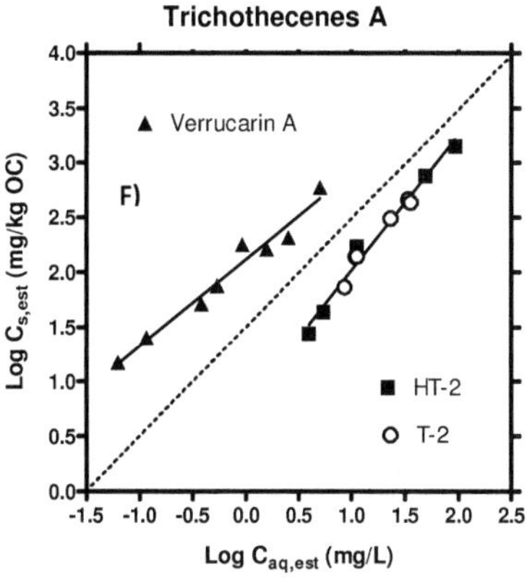
Trichothecenes A
F)
Verrucarin A
HT-2
T-2
Log Cs,est (mg/kg OC)
Log Caq,est (mg/L)

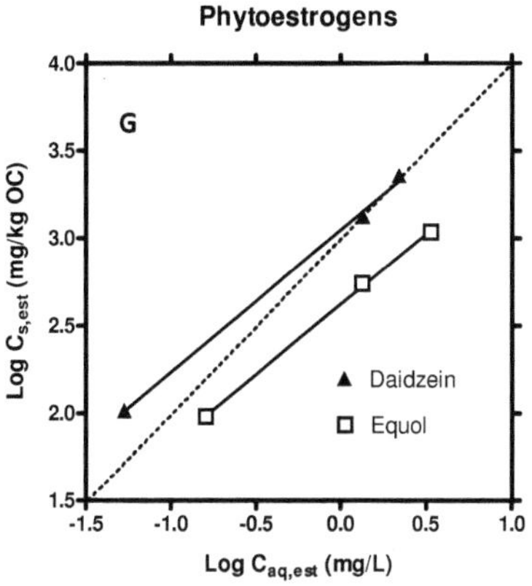
Phytoestrogens
G
Daidzein
Equol
Log Cs,est (mg/kg OC)
Log Caq,est (mg/L)

Table S4.4: Sorption data details including the tested concentration [mg/L and mg/kg OC], tested column, Freundlich constant (K_{OC} at 1 mg/L) and exponent and overall fitting error, K_{OC} at 100 mg/kg$_{oc}$.

COMPOUND	tested range $C_{W, est}$ [mg/kg]	tested range $C_{S, est}$ [mg/kg]	columns (% OM)	*N*	Log K_{oc} at 1 mg/L [c]	Freundlich exponent [c]	sy.x [d]	Log K_{oc} at 100 mg/kg$_{oc}$
Ergot alkaloids								
Ergocornine	0.01 – 1.2	76 - 4480	3%	5	3.6 ±	0.83 ± 0.02	0.032	3.88
Ergocryptine	0.02 – 0.3	150 – 1400	1%	4	3.5 ±	0.75 ± 0.08	0.075	4.02
Penicillium								
Citrinin	0.2 – 4.4	250 – 2800	1.5%	3	2.9 ±	0.83 ± 0.02	0.014	3.10
Patulin	0.1 – 1.5	2.5 – 25	10% x 5cm	3	1.2 ±	0.82 ± 0.03	0.028	1.08
Aflatoxins								
Aflatoxin M_1	0.7 – 6.1	570 - 2730	1%	3	2.9 ±	0.73 ± 0.04	0.028	3.18
Aflatoxin B_1	0.1 – 11	127 - 9600	1%-1.5%	12	3.2 ±	0.80 ± 0.02	0.039	3.46
Aflatoxin B_2	0.4 - 3.3	340 - 1870	1%	3	2.9 ±	0.82 ± 0.04	0.025	3.05
Aflatoxin G_1	0.7 – 4.3	710 - 2670	1%	3	3.0 ±	0.74 ± 0.01	0.003	3.31
Aflatoxin G_2	0.2 – 3.6	110 - 1070	1%	3	2.6 ±	0.78 ± 0.04	0.037	2.80
Alternaria toxins								
Alternariol MME	0.16 – 0.28	5.5 - 8.1	5%	3	1.3 ±	0.70 ± 0.04	0.007	1.00
Alternariol	0.06 – 0.6	8.2 - 61	1%-1.5%	6	2.1 ±	0.88 ± 0.17	0.131	2.10
Altenuene	0.02 – 1.2	8.5 - 450	3%	12	2.6 ±	0.97 ± 0.04	0.073	2.58
Tentoxin	0.2-1.9	3.8 - 69	10% x 5cm	3	1.4 ±	0.97 ± 0.06	0.054	1.39
Resorcylic acid lactones								
Zearalenone	0.02 - 3.2	53 - 3790	1%-3%	8	3.3 ±	0.90 ± 0.03	0.040	3.42
α-Zearalenol	0.01 – 0.9	11 – 720	1%-3%	10	2.8 ±	0.90 ± 0.05	0.105	2.87
β-Zearalenol	0.01 – 0.5	7.6 - 320	1%-3%	13	2.7 ±	0.93 ± 0.06	0.106	2.81
Trichothecenes A								
T-2 Toxin	8.6 – 35	73 – 460	10% x 5cm	6	0.8 ±	1.17 ± 0.11	0.068	1.03
HT-2 Toxin	4 – 93	28 - 1410	10% x 5cm	5	0.8 ±	1.24 ± 0.10	0.113	1.01
Neosolaniol		n.d. [b]	10% x 5cm					n.d. [b]
Diacetoxyscirpenol		n.d.	10% x 5cm					n.d.
Verrucarol		n.d.	10% x 5cm					n.d.
Verrucarin A	0.06 – 5.1	25 - 588	3%	8	2.2 ±	0.79 ± 0.06	0.101	2.15
Trichothecenes B								
3-Acetyl-DON		n.d.	10% x 5cm					n.d.
15-Acetyl-DON		n.d.	10% x 5cm					n.d.
Deoxynivalenol		n.d.	10% x 5cm					n.d.
Fusarenone-X		n.d.	10% x 5cm					n.d.
Nivalenol		n.d.	10% x 5cm					n.d.
Other mycotoxins								
Beauvericin		n.a.[b]						n.a.
Sulochrin	0.4 - 3.1	85 - 1214	5%	3	1.9 ±	1.00 ± 0.03	0.016	1.93
Sterigmatocystin	0.5 – 1.9	38 - 405	1%	3				2.07 [a]
Phytoestrogens								
Daidzein	0.1 - 2.2	103 - 2272	1%	3	3.0 ±	0.81 ± 0.03	0.044	3.29
Equol	0.2 - 3.4	96 - 1080	1%	3	2.6 ±	0.80 ± 0.02	0.022	2.78

[a] taken from average sorption coefficient from all tested data points because sorption isotherm slope >1.1; [b] n.a.=not analyzed because retention to column with only SiC was too high, n.d.=not detected because net retention volume < 30% above net retention volume of non-retaining tracer; [c] fitted Freundlich parameters with data in mg/L for estimated dissolved concentration (from peak maximum, $C_{AQ, est}$) and estimated sorbed concentrations in mg/kg OC (from $C_{AQ, est}$ and retention time K_{OC}); [d] sy.x = overall error margins of fitted curve.

Table S4.5. Experimental log K_{oc} values (normalized to 100 mg/kg$_{oc}$) and predicted log K_{oc} using different models.

Compound	Column method Pahokee peat	EPIsuite v4.0 PCKOCNT v2.00		Sparc v4.5 NOM model from[8]	COSMO*therm* NOM model from[8]	pp-LFER (LSABV) eq 6[1]
Neutral mycotoxins	**exp. Log K_{OC}***	**via MCI**	**via K_{OW}**			
Penicillium						
Patulin	1.08	-0.14	-0.97	2.45	0.64	-0.69
Aflatoxins						
Aflatoxin M_1	3.18	0.64	0.52	5.67	2.33	0.08
Aflatoxin B_1	3.46	1.78	1.76	5.22	2.58	1.10
Aflatoxin B_2	3.05	1.78	1.88	5.03	2.08	1.93
Aflatoxin G_1	3.31	1.87	1.09	4.71	2.31	0.27
Aflatoxin G_2	2.80	1.87	1.21	4.75	1.78	0.65
Alternaria toxins						
Alternariol MME	1.00	3.39	2.86	4.80	3.50	3.06
Alternariol	2.10	3.69	2.66	4.72	2.23	2.28
Altenuene	2.58	1.03	1.49	6.01	2.37	1.92
Tentoxin	1.39	4.63	1.59	10.33	2.19	0.09
Resorcylic acid lactones						
Zearalenone	3.42	3.70	3.37	4.86	3.44	3.61
α-Zearalenol	2.87	3.51	3.04	6.29	3.67	3.90
β-Zearalenol	2.81	3.51	3.05	6.29	3.64	3.90
Trichothecene A						
T-2 Toxin	1.03	2.96	1.46	8.29	3.57	2.59
HT-2 Toxin	1.01	0.92	0.71	8.19	2.64	2.18
Neosolaniol	<0.7	0.19	-0.06	7.24	2.19	0.91
Diacetoxyscirpenol	<0.7	1.29	0.98	6.47	2.69	1.83
Verrucarol	<0.7	0.91	0.19	6.23	1.99	1.86
Verrucarin A	2.15	3.87	-1.64	8.33	3.89	3.49
Trichothecene B						
3-Acetyl-DON	<0.7	-0.35	-0.25	6.15	0.93	0.36
15-Acetyl-DON	<0.7	-0.37	0.22	6.15	1.01	0.17
Deoxynivalenol	<0.7	0.22	-0.28	6.58	0.84	-0.41
Fusarenone-X	<0.7	-0.12	-0.64	7.24	2.42	-0.84
Nivalenol	<0.7	0.45	-1.12	7.64	0.70	-1.26
Other mycotoxins						
Beauvericin	n.a.	8.75	4.39	14.46	n.a.	5.90
Sulochrin	1.93	3.09	4.03	5.93	3.23	2.35
Sterigmatocystin	2.07	3.24	3.42	5.90	4.53	2.81
Phytoestrogens						
Daidzein	3.29	3.37	2.92	4.46	1.95	0.84
Equol	2.78	4.27	3.35	5.23	2.78	2.27
Ionized mycotoxins						
Ergocornine	3.88					
Ergocryptine	4.02					
Citrinin (96%-)	3.10					

* Log K_{OC} was derived for a sorbed concentration of 100mg/kg$_{OC}$, the variation of the experimental Log K_{OC} value was determined separately and is estimated to be well within 0.2 log units. n.a. is not analyzed.

Section SI-4.4 Description of predictive tools used for the determination of K_{ow} and K_{oc} values for the natural toxins

For the single-parameter approach to derive K_{oc} from K_{ow} values, no measured K_{ow} values are available for any of the mycotoxins, only for the two isoflavones are K_{ow} values reported.[2] Therefore, various tools to estimate K_{ow} values were assessed;

(i) KowWin program from U.S. EPA's EPISuite 4.1 package (http://www.epa.gov/opptintr/ exposure/pubs/episuitedl.htm), online tool SPARC v4.5 (http://archemcalc.com/sparc/) and commercial software ALOGPS 2.1 (http://www.vcclab.org/ lab/alogps/).

(ii) Non-optimized .mol-files of the neutral compounds were converted from SMILES format using online converter (http://www.webqc.org/ molecularformatsconverter.php), and used as input for the three methods available in online tool Marvinsketch (http://www.webqc.org/moleculareditor2.php; VG-method (incremental method by Viswanadhan et al.[9]), KLOP-method (LogP data from Klopman et al.[10]) and PHYS-method (using LogP data from PHYSPROP© database)).

(iii) input files for *COSMOtherm*[11] software calculations (version C21_0111, COSMOlogic GmbH & Co. KG, Leverkusen, Germany, 2010) were high quality geometrically optimized molecules using the program TURBOMOLE (version 6.0, TURBOMOLE, a development of the University of Karlsruhe and Forschungszentrum Karlsruhe GmbH, 1989-2007, TURBOMOLE GmbH, since 2007 available from www.turbomole.com). *COSMOtherm* calculation on partitioning coefficients are done using the COSMO-RS method, a combination of the quantum-chemical dielectric continuum solvation model (COSMO) with a statistical thermodynamics treatment of surface interactions (COSMO-RS). In order to create the input structures for *COSMOtherm*, for each test compound calculations were started with obtaining lowest-energy 3D conformations (performed with COSMOconf, version 2.1), followed full geometry optimization in the gas phase and in the conductor reference state

(COSMO) by BP-TZVP gas-phase and COSMO calculations with the TURBOMOLE program. The gas-phase energies and COSMO files of the sorbates that resulted from the TURBOMOLE calculations were then used for further calculations in the *COSMOtherm* software.

Direct K_{oc} estimates were obtained from (i) KOCWIN v2.00 software within EPISuite Package V4.1 using MCI-method and K_{ow}-method, (ii) SPARC v.4.5 by calculating the distribution coefficient between water and the NOM model presented in Niederer et al.[8] for all mycotoxins, and for five other NOM models for humic acids described in Atalay et al.[12] for three mycotoxins and several simple chemical structures, (iii) *COSMOtherm* v2.1 using the same NOM model but then as fully geometrically optimized neutral NOM structure, (iv) the calibrated pp-LFER equation for micronized Pahokee peat from Bronner & Goss.[1] Experimental pp-LFER descriptors are only available for zearalenone in the currently studied set of mycotoxins and isoflavones. The pp-LFER sorbate descriptors can also be estimated from molecular structure with the program ABSOLV, Algorithms ADME Boxes (http://www.acdlabs.com/products/pc_admet/).[13] The K_{oc} pp-LFER defined by Bronner & Goss,[1] however, is calibrated only with experimentally derived descriptors.

Section SI-4.5 Comparison experimental K_{oc} with predicted log K_{ow} and predicted log K_{oc}

Figure S4.6. Log K_{ow} values for neutral mycotoxins predicted with various tools. The figure on the left shows a boxplot of predicted log K_{ow} values with 25th and 75th percentile, the vertical bar depicting the average, and horizontal bars stretching from minimum to maximum values. The figure on the right show the same data set, but now plotted as individual predictive values, with EPISuite KowWIN values depicted as X, and COSMOtherm values depicted as large grey dots. Other K_{ow} estimates are from SPARC v.4.5, ACDLabs v.12, ALOGPs v.2.1 and three LogP algorithms available via online tool MarvinSketch. Indicated with red squares are the experimental K_{oc} data. Note that for those compounds where sorption was lower than the dynamic retardation factor, where Log K_{oc} <0.7, data are indicated with red triangles.

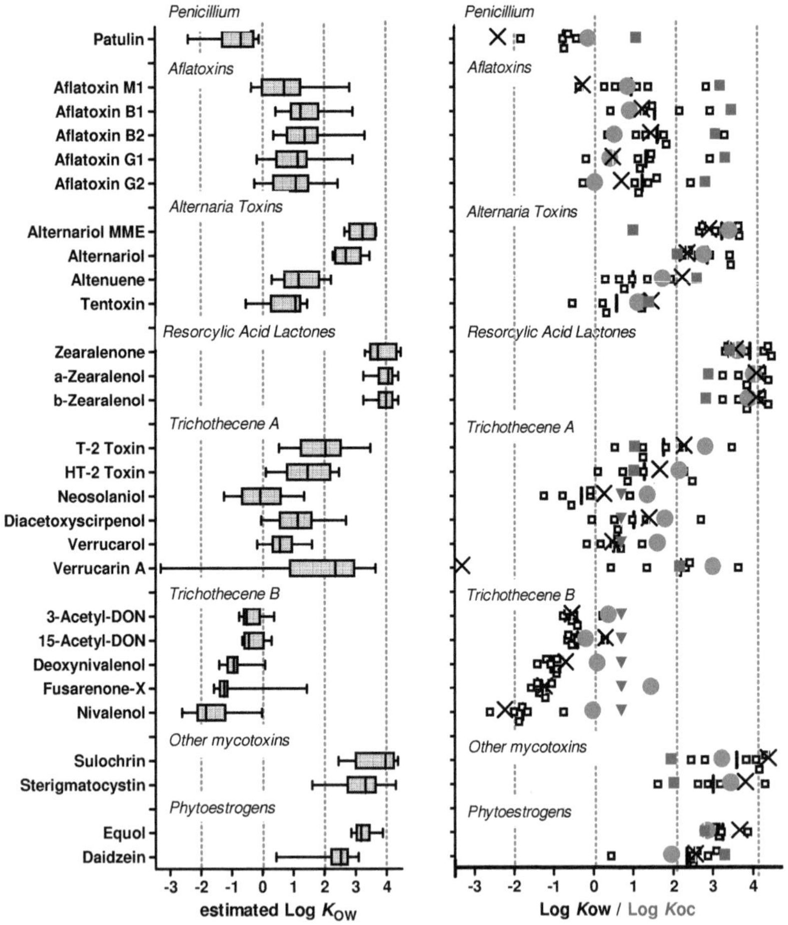

Figure S4.7. **A)** Comparison for a diverse set of polar compounds from[4] of experimental log K_{ow} versus predicted log K_{ow} using ALOGPS 2.1; **B)** Comparison of experimental determined log K_{oc} values and log K_{ow} values for a diverse set of polar compounds from;[1,4,5] **C)** Experimental K_{oc} data for mycotoxins (at 100 mg/kg$_{oc}$) plotted against predicted log K_{ow} values obtained with ALOGPS, SPARC, *COSMOtherm* and pp-LFER with ABSOLV estimated parameters from **Table S4.3**.

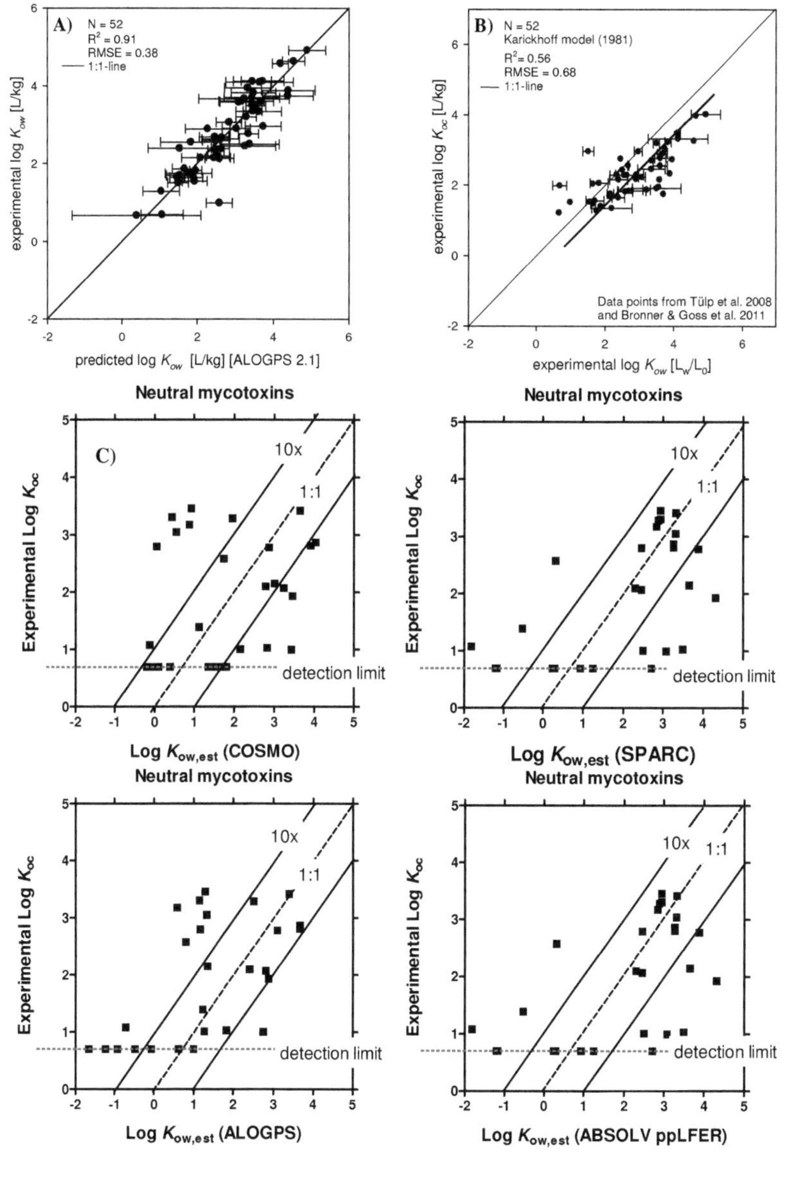

Figure S4.8. Experimental log K_{oc} sorption data (normalized at 100 mg/kg$_{oc}$) for neutral or the molecular connectivity index method (**MCI; B**).

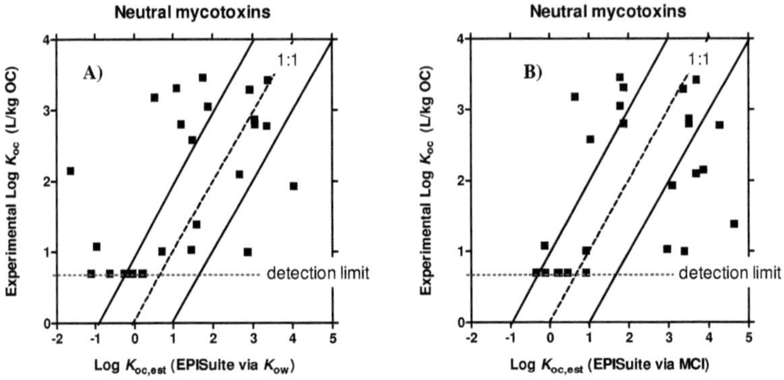

Figure S4.9. Experimental log K_{oc} sorption data (normalized at 100 mg/kg$_{oc}$) for neutral mycotoxins plotted against predicted log K_{oc} values by pp-LFER equation for Pahokee peat,[1] applying estimated pp-LFER descriptors obtained with ABSOLV software.

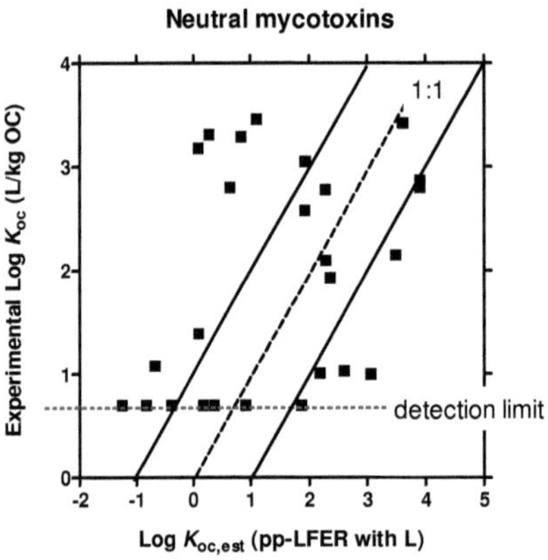

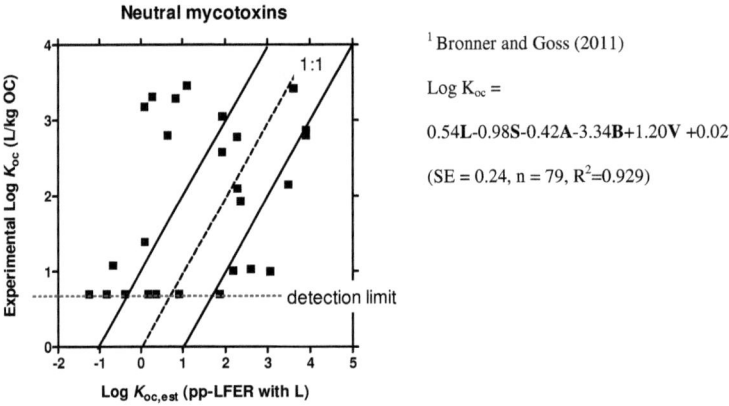

[1] Bronner and Goss (2011)

Log K_{oc} =

0.54**L**-0.98**S**-0.42**A**-3.34**B**+1.20**V** +0.02

(SE = 0.24, n = 79, R^2=0.929)

Figure S4.10. SPARC predicted K_{oc} sorption data for various NOM models, SMILES format taken from ref.[14] **Top)** neutral mycotoxins (zearalenone, HT-2 toxin and patulin) **Lower Graph)** simple neutral structures.

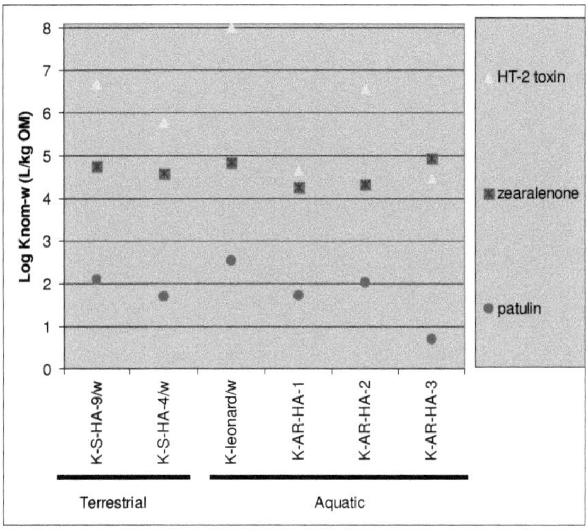

compound	S-HA-9	S-HA-4	LEO	AR-HA-1	AR-HA-2	AR-HA-3	span LogK
HT-2 toxin	6.7	5.8	8.0	4.6	6.6	4.5	**3.6**
zearalenone	4.8	4.6	4.9	4.3	4.3	4.9	**0.7**
patulin	2.1	1.7	2.5	1.7	2.0	0.7	**1.9**

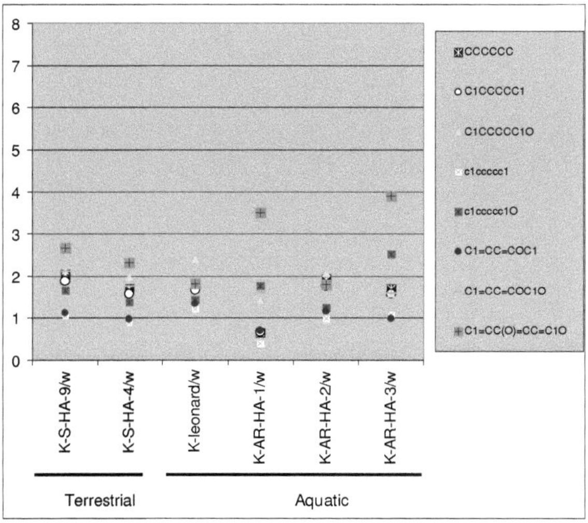

	S-HA-9	**S-HA-4**	**LEO**	**AR-HA-1**	**AR-HA-2**	**AR-HA-3**	**span LogK**
CCCCCC	2.0	1.7	1.8	0.7	1.9	1.7	**1.4**
C1CCCCC1	1.9	1.6	1.7	0.7	1.8	1.6	**1.2**
C1CCCCC1O	2.1	2.0	2.4	1.4	2.1	1.6	**1.0**
c1ccccc1	1.1	0.9	1.2	0.4	1.0	1.1	**0.9**
c1ccccc1O	1.7	1.4	1.4	1.8	1.2	2.5	**1.3**
C1=CC=COC1	1.1	1.0	1.3	0.7	1.2	1.0	**0.7**
C1=CC=COC1O	2.1	1.8	1.9	1.6	1.9	1.6	**0.5**
C1=CC(O)=CC=C1O	2.7	2.3	1.8	3.5	1.8	3.9	**2.1**

NOM	**composition**	**OC %**	**SPARC Unique SMILES structure**
S-HA-9	$C_{45}H_{45}NO_{23}S$	54%	O=C(O)c5cc(NC(C(=O)O)CCSC)c(C(=O)O)c(Oc4cc(Oc3cc(O)cc(OC2(O)(C(O)-C(OCc1cc(OC)ccc1CC(=O)O)OCC2(C(=O)O)))c3)ccc4CCC(=O)O)c5(C(=O)O)
S-HA-4	$C_{44}H_{41}NO_{26}S$	51%	O=C(O)c5ccc(OC)c(Oc4ccc(c(OC3OC(Oc2c(C(=O)O)c(Oc1ccc(O)c(c1)NC(C(=O)O)CCSC)cc(C(=O)O)c2(C(=O)O))(C(=O)O)C(O)(O)C(C(=O)O)C3(O))c4)CCC(=O)O)c5
leonardite	$C_{31}H_{26}O_{12}$	63%	O=C(O)c4cc(Oc1c(OC)cc(cc1(C(=O)O))C(C2=CC(=O)c3cc(C(=O)O)c(OC)cc3-(C2(=O)))CCC)ccc4(O)
AR-HA-1	$C_{38}H_{34}O_{22}$	54%	O=C(O)C=Cc4ccc(Oc1ccc(cc1(OC))C(=O)Oc2c(O)cc(cc2(O))C(=O)OC(C(O)C-(O)O)C(O)C(O)C(=O)Oc3c(O)cc(cc3(O))C(=O)O)c(OC)c4
AR-HA-2	$C_{43}H_{43}NO_{22}$	56%	O=C(O)c5cc(OC(=O)C(NC(=O)C(O)(OC)C(=O)Oc1ccc(cc1(OC))C2OCC(OC2)-c3ccc(O)c(OC)c3)CCC(=O)O)c(O)c(Oc4c(OC)cc(cc4(OC))C(=O)O)c5
AR-HA-3	$C_{31}H_{42}N_2O_{14}$	56%	O=C(O)c1ccc(cc1)C(NCCCCCCC(O)OC3OC(CO)C(Oc2c(O)cc(cc2(O))C(=O)-O)C(O)C3(NC(=O)C))CO

Figure S4.11. Experimental log K_{oc} sorption data (normalized at 100 mg/kg$_{oc}$) for neutral mycotoxins plotted against predicted log K_{oc} values for 'leonardite' NOM model by SPARC (**Left**) and COSMOtherm (**Right**). SPARC uses the SMILES format of the NOM structure as input for calculations, whereas COSMOtherm uses an estimated charge density on a high quality optimization of the 3D structure.

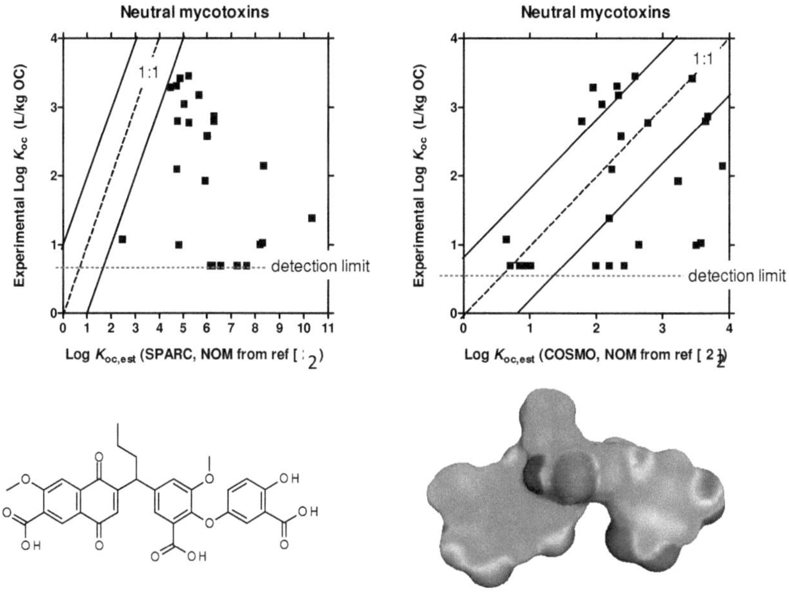

References (chapter 4 supporting information):

1. Bronner, G. and Goss, K.U., Predicting sorption of pesticides and other multifunctional organic chemicals to soil organic carbon. *Environ. Sci. Technol.* **2011,** *45*, 1313-1319.
2. Rothwell, J.A., et al., Experimental determination of octanol-water partition coefficients of quercetin and related flavonoids. *J. Agric. Food Chem.* **2005,** *53*, 4355-4360.
3. Abraham, M.H., et al., Hydrogen bonding. 32. An analysis of water-octanol and water-alkane partitioning and the Dlog P parameter of seiler. *J. Pharm. Sci.* **1994,** *83*, 1085-1100.

4. Tulp, H.C., et al., Experimental determination of LSER parameters for a set of 76 diverse pesticides and pharmaceuticals. *Environ. Sci. Technol.* **2008,** *42*, 2034-2040.

5. Bronner, G. and Goss, K.U., Sorption of organic chemicals to soil organic matter: influence of soil variability and pH dependence. *Environ. Sci. Technol.* **2011,** *45*, 1307-1312.

6. Tulp, H.C., et al., pH-Dependent sorption of acidic organic chemicals to soil organic matter. *Environ. Sci. Technol.* **2009,** *43*, 9189-9195.

7. Gustafsson, Ö. and Bucheli, T., Soot sorption of non-ortho and ortho substituted PCBs *Chemosphere* **2003,** *53*, 515-522.

8. Niederer, C. and Goss, K.U., Quantum chemical modeling of humic acid/air equilibrium partitioning of organic vapors. *Environ. Sci. Technol.* **2007,** *41*, 3646-3652.

9. Viswanadhan, V.N., et al., Atomic Physicochemical Parameters for 3 Dimensional Structure Directed Quantitative Structure - Activity Relationships .4. Additional Parameters for Hydrophobic and Dispersive Interactions and Their Application for an Automated Superposition of Certain Naturally-Occurring Nucleoside Antibiotics. *Journal of Chemical Information and Computer Sciences* **1989,** *29*, 163-172.

10. Klopman, G., et al., Computer Automated Log P Calculations Based on an Extended Group-Contribution Approach. *Journal of Chemical Information and Computer Sciences* **1994,** *34*, 752-781.

11. Eckert, F. and Klamt, A., *COSMOtherm*, Version C2.1; COSMOlogic GmbH & Co. KG, Leverkusen, Germany: 2009.

12. Atalay, Y.B., et al., Distribution of Proton Dissociation Constants for Model Humic and Fulvic Acid Molecules. *Environ. Sci. Technol.* **2009,** *43*, 3626-3631.

13. ABSOLV ACD/ADME Suite - Advanced Chemistry Development. Toronto, Ontario, Canada

14. Phillips, K.L., et al., Prediction of soil sorption coefficients using model molecular structures for organic matter and the quantum mechanical *COSMO*-SAC model. *Environ. Sci. Technol.* **2011,** *45*, 1021-1027.

CHAPTER 5:

MYCOTOXINS IN THE ENVIRONMENT: I. PRODUCTION AND EMISSION FROM AN AGRICULTURAL TEST FIELD

JUDITH SCHENZEL [1,2], HANS-RUDOLF FORRER[1], SUSANNE VOGELGSANG[1], KONRAD HUNGERBÜHLER[2], THOMAS D. BUCHELI[1]

[1]AGROSCOPE RECKENHOLZ-TANIKON, RESEARCH STATION ART, RECKENHOLZSTRASSE 191, CH-8046 ZURICH, SWITZERLAND
[2]INSTITUTE OF CHEMICAL AND BIOENGINEERING, SWISS FEDERAL INSTITUTE OF TECHNOLOGY, WOLFGANG-PAULI STRASSE 10, CH-8093 ZURICH, SWITZERLAND

REPRODUCED WITH PERMISSION FROM:

ENVIRONMENTAL SCIENCE AND TECHNOLOGY, 2012

VOLUME 46, ISSUE 24, PAGES 13067-13075

Abstract

Mycotoxins are secondary metabolites that are naturally produced by fungi which infest and contaminate agricultural crops and commodities (e.g., small grain cereals, fruits, vegetables, and organic soil material). Although these compounds have extensively been studied in food and feed, only little is known about their environmental fate. Therefore, we investigated over nearly two years the occurrence of various mycotoxins in a field cropped with winter wheat of the variety Levis, which was artificially inoculated with *Fusarium* spp., as well as their emission via drainage water. Mycotoxins were regularly quantified in whole wheat plants (0.1–133 $mg/kg_{dry\ weight}$, for deoxynivalenol), and drainage water samples (0.8 ng/L – 1.14 µg/L, for deoxynivalenol). From the mycotoxins quantified in wheat (3-acetyl-deoxynivalenol, deoxynivalenol, fusarenone-X, nivalenol, HT-2 toxin, T-2 toxin, beauvericin, and zearalenone), only the more hydrophilic ones or those prevailing at high concentrations were detected in drainage water. Of the total amounts produced in wheat plants (min: 2.3; max: 292 g/ha/y), 0.5–354 mg/ha/y, i.e. 0.002–0.12%, were emitted via drainage water. Hence, these compounds add to the complex mixture of natural and anthropogenic micropollutants particularly in small rural water bodies, receiving mainly runoff from agricultural areas.

5.1 Introduction

During cultivation, agricultural commodities are at risk of infection from a number of different fungi, and of contamination with their toxic metabolites, called mycotoxins. These fungi include species of *Fusarium, Alternaria, Cladosporium,* and *Claviceps*. Especially, crops like small grain cereals and maize could contain a number of potential mycotoxins at harvest. Thus far, several hundred different mycotoxins have been discovered, exhibiting great structural diversity, which results in widely varying chemical and physicochemical properties.[1] Fumonisins, trichothecenes and zearalenone (produced by *Fusarium* species), and ergot alkaloids (produced by *Claviceps* species) are most prominent due to their frequent occurrence and their severe effects on animal and human health.[2,3] Mycotoxins have been ranked as the most important chronic dietary risk factor, higher than synthetic contaminants, plant toxins, food additives or pesticide residues.[4] Thus, much research has been carried out on the analysis of mycotoxins in food and feed matrices, human and husbandry animal exposure, and related health effects.[3,5,6]

However, the environmental exposure to mycotoxins has been scarcely investigated. Recent studies on the occurrence of two *Fusarium* produced mycotoxins, deoxynivalenol (DON) and the estrogenic zearalenone (ZON), in plants, soil, and drainage water of an infected wheat and maize field reported concentrations in drainage water and corresponding emission rates of 20–5000 ng/L and 600 mg/ha for DON, and up to 35 ng/L and 3 mg/ha for ZON,[7-9] resulting in a frequent (DON) and occasional (ZON) detection of these two mycotoxins in river waters.[9,10] These results demonstrated that the run-off from agricultural fields is one significant source of mycotoxins in surface waters. Other mycotoxins, like nivalenol (NIV), or 3-acetyl-deoxynivalenol (3-AcDON), are produced in similar amounts as DON and ZON, and exhibit similar or even higher water solubility than DON. Therefore, it is likely that further mycotoxins are introduced into the aqueous environment. In fact, Schenzel et al.[11] observed three more mycotoxins, i.e. 3-AcDON, NIV, and beauvericin (BEA), in the drainage water from a wheat field by using a multi-residue screening method for various classes of mycotoxins.

The aim of this study was to investigate systematically and over two growing seasons the production of a diverse set of mycotoxins on agricultural plots cropped with wheat, and to quantify the amounts emitted via drainage water discharge. Therefore, an experimental field was repeatedly cultivated with winter wheat of the variety Levis. To assure disease and mycotoxin production the crop was artificially infected with four *Fusarium* species (*F. avenaceum*, *F. crookwellense*, *F. graminearum*, *F. poae*) commonly found in wheat. For almost two years, mycotoxins were monitored regularly in whole wheat samples, and continuously in drainage water. To the best of our knowledge, this is the first time that such an approach is used to examine the risk of environmental exposure to several classes of mycotoxins. Additionally, the ecotoxicological relevance of the presence of these compounds in drainage waters will be briefly evaluated. Finally, in a companion publication, the exposure of river water by mycotoxins is presented in detail, and the relative contribution of diffuse (i.e. *Fusarium* infected wheat crop[8,9]), and point sources (i.e. waste water treatment plants[12,13]) is assessed.

5.2 Materials and Methods

5.2.1 Field site description, instrumentation, and cultivation

Wheat cultivation and drainage water sampling took place on two fields located near to the research station Agroscope ART Reckenholz, North of Zurich, Switzerland (47°25`74" N, 8°30`85" E). The two adjacent but separately drained fields of 0.2 hectare exhibit a gentle slope of 1–2°. The top soil of both of these test sites was classified according to the World reference base for soil resources and shows characteristics of a medium to heavy textured gleyic cambisol with 31% clay, 30% sand, 39% silt, and an organic carbon fraction of 2%.[14] A detailed list of all soil parameters is given in the supporting information (**Table S5.1**). Both test fields were drained by two long and two short drainage tubes, which individually connect to a main drainage tube with a diameter of 15 cm (**Figure S5.1**). The drainage tubes are located in a depth of 80 to 90 cm. The ground water table depth ranged from 100–125 cm and thus was permanently below the drainage system.[8] Both drainage tubes ended in sampling ducts, which were equipped with flow meters and automated samplers (7612 ISCO with a 730 bubbler module, both from Teledyne Isco Inc. Lincoln, USA) for flow proportional sampling of discharged drainage water. Precipitation data were gathered by the meteorological station (Reckenholz 443 m above sea level, 47°25`40" N, 8°31`04" E, MeteoSwiss, approx. 300 m from the field site) in 10 min intervals and are depicted in **Figure S5.2** in the **SI**.

To study the potential production and run-off of mycotoxins from an agricultural field, the following conditions favourable for mycotoxin producing *Fusarium* species infestation[15] were chosen for the upper part of the test field (**Figure S5.1**,I): (i) winter wheat (*Triticum aestivum* L.) was cultivated repeatedly over the entire investigation period from 2008 until 2011, (ii) a wheat variety (Levis), which is highly susceptible to *Fusarium graminearum* infestation, was selected, (iii) no soil cultivation ("no-till") or a superficial tillage with a temporary removal of the straw was performed and wheat straw was spread back onto the surface of the original field plots, and (iv) during flowering in 2009, 2010, and 2011 the winter wheat was artificially infected with four different *Fusarium* species including *F. avenaceum, F. crookwellense, F. graminearum,* and *F. poae.* The fields were cultivated using standard farming conditions, except that no fungicides were applied. Sowing dates for wheat where October 29th, and October 12th in 2009, and in 2010, respectively.

In particular, inoculations of the upper field were carried out with the different *Fusarium* species spore suspensions applied individually on four different subplots of equal size (11x22 m; space between plots ≥0.5 m, **Figure S5.2**). The four different species comprised the following strains isolated by Agroscope ART from different sites in Switzerland (all deposited at the Centraalbureau voor Schimmelcultures, Utrecht, NL): *F. avenaceum* CBS 121289, CBS 121294, and CBS 121290; *F. crookwellense* CBS 121293; *F. graminearum* CBS 121291, CBS 121292, and CBS 121296; *F. poae* CBS 121298, CBS 121297, and CBS 121299. Fungal suspensions (equal amounts of each strain) were prepared with the nonionic detergent Tween® 20 (Promega AG, Dübendorf, Switzerland) at 0.0125% to obtain final concentrations of 2.5×10^5 conidia/mL, which were subsequently applied on each subplot. The control subplot received a treatment with distilled water (Tween at 0.0125%) only. In the control plot only the slightly contaminated straw from the control subplot itself was laid out again after the harvest, leading to a potential re-infection of the wheat plants with *Fusarium* species. In all subplots, the final suspensions were applied at a total volume of 730 L/ha using a back-pack sprayer (width 1.5 m, 3 bar, Birchmeier M125, Birchmeier Sprühtechnik AG, Stetten, Switzerland) covering the entire plot surface. Please note that the chosen experimental setup was not designed to investigate and statistically evaluate the relationship between Fusarium *spp.* infection, chemotypes and agricultural and climatic influences (for this widely investigated field of research, we refer to the specialized literature, e.g.,[16-20]), but to assure the presence of a wide variety of mycotoxins, and to obtain a prominent input signal of these compounds into the agro-environment, allowing studying their emission and fate via drainage water.

The lower field (**Figure S5.1**,II) was used in 2011 only. In contrast to the upper part, the whole field was ploughed, and the seedbed was prepared in early march using a rotary harrow. Afterwards, spring wheat (*Triticum aestivum* L.) of the variety Rubli was planted. Here, no artificial infection was conducted, and no wheat straw was laid out on the lower field. Therefore, this field serves as an additional control treatment.

A seed health test (SHT) was conducted to determine which species and to which extent the species was responsible for an incidence of *Fusarium* spp. infection.[21] Briefly, grains were surface-sterilized (10 min, 1% chloramine T) and 100 grains were placed on potato dextrose agar (PDA) (10 grains/plate) and incubated for 6 days at 19±1°C with a photoperiod of 12 h dark/12 h near-UV light. Colonies of *Fusarium* species and *M. nivale*/majus were identified according to Leslie and Summerell.[22] Fungal incidences in each sample were calculated in percentage.

5.2.2. Field sampling, sampling preparation and extraction

The wheat plants from the individual subplots of the upper field were sampled on July 29, 2010, i.e. immediately before the harvest, and at two times from mid of June 2011 until harvest at July 16, in 2011. Each sample consisted of five to ten wheat plants that were collected manually from just above the ground level from randomly selected locations of a given subplot. Sampling in the lower field in 2011 was conducted concomitant with the upper one, and in the same manner from the whole field. Samples were processed and analysed as described in Schenzel et al.[23] Briefly, collected wheat plant samples including stems, leaves and entire ears were freeze-dried, ground and homogenized prior to extraction. The whole-wheat plant samples were extracted using a simple solid-liquid extraction approach with 20 mL of MeCN/Milli-Q water (80:20, v/v) for 2 h on a SM 30 orbital shaker at 200 rpm (Edmund Bühler GmbH, Hechingen, Germany). The extract was filtered and further applied to a clean-up procedure using a Varian Bond Elut Mycotoxin® cartridge. Finally, 1 mL of the extract was reduced to 100±5 μL, further reconstituted in 900 μL of Milli-Q water/MeOH (90:10, v/v), and transferred into 1.5-mL amber glass vials. The samples were stored at +4°C and analyzed within 48 h after extraction.

Drainage water samples of the upper and lower field were gathered from December 1, 2009 until October 31, 2011. They were taken flow proportionally at a sampling rate of 200 mL per 1000 L. Five subsamples were combined to one. This type of sampling allows to quantify pollutant loads in a given time period from measured concentrations and actual water flow. Drainage water samples were processed and analysed as described in Schenzel et al.[11] Briefly, filtered water samples (1 L) were concentrated and purified by performing reversed-phase solid-phase extraction (Oasis HLB cartridges, 6 mL, 200 mg; Waters Corporation, Milford MA, USA). Analytes were eluted with 5 mL of MeOH, further reduced to 100±5 μL, reconstituted in 900 μL of Milli-Q water/MeOH (90:10, v/v), and transferred into 1.5-mL amber glass vials. The samples were stored at +4°C and analyzed within 48 h. Details about both analytical methods and their quality control are also given in the **SI**.

5.2.3. Analysis and data presentation

All mycotoxins were quantified as described previously.[11,23] For a complete list of compounds and their selection criteria, we refer to this original literature. Briefly, LC-MS/MS was performed on a Varian 1200L LC-MS instrument (VarianInc, Walnut Creek CA, USA). The mycotoxins were separated using a 125 mm × 2.0 mm i.d., 3 µm, Pyramid C_{18} column, with a 20 mm x 2.0 mm i.d. 3 µm guard column of the same material (Macherey-Nagel GmbH & Co. KG, Düren, Germany). Two runs with varying mobile phase gradient and different ionization mode (pos, neg) were used for the detection of all compounds. Further details about the LC-MS method are also given in the **SI**. If an analyte was detectable but not quantifiable, its concentration was set equal to its limit of detection (LOD). LODs of the 28 mycotoxins ranged from 1 to 26 µg/kg in whole wheat plants[11], and from 0.3 to 10.7 ng/L in drainage water.[23] Mycotoxin loads in the drainage water were calculated from the quantified concentrations and the drainage water discharge, which was obtained from the ISCO sampler protocols.

5.3 Results and discussion

Mycotoxins produced by *Fusarium* species prevailed both in whole wheat plants and in drainage water. As expected, no aflatoxins, ochratoxins, nor *Penicillium* toxins, which are commonly known to be mainly produced during storage,[24] were detected, neither in plant nor in drainage water samples. Over the entire period of investigation, only eight, and five out of 28 and 33 mycotoxins, which were included in the analytical methods for solid and aqueous samples, respectively, were repeatedly detected (**Table 5.1** and **5.2**). The remainder of this paper will therefore focus on these.

5.3.1 Mycotoxins in whole wheat plants

Artificial infection of the upper wheat field's individual subplots (**Figure S5.1**,I) in June 14, 2010 and May 26, 2011 led to highly varying *Fusarium* species incidence ranging from five (control plot) up to 103% (*F. graminearum* subfield) for the sum of all *Fusarium* species (**Table S5.6**). Generally, seeds were mostly contaminated with the *Fusarium* species they were actually inoculated with, but the sum of all infections can be slightly higher than 100% due to some cross-contamination (**Table 5.1 Table S5.6**). Mycotoxin concentrations and patterns in wheat plants varied both over time and in between subfields (**Table 5.1**). In addition, the absolute amount of mycotoxins at harvest differed from plot to plot, as well as in between both years. Although largely deliberately provoked by artificial *Fusarium* infection, such variability is similar to the one taking place under natural conditions.[25] Within all subfields and in both years of investigation, DON was by far the compound which was quantified with the highest concentrations of up to 133 mg/kg_{dw}, followed by NIV (<12 mg/kg_{dw}), ZON (<7.4 mg/kg_{dw}), 3-AcDON (<3.1 mg/kg_{dw}), and BEA (<2.2 mg/kg_{dw}). Again, while such concentrations certainly are high, they are not completely beyond what can be observed in practice.[26,27]

In 2010, a *Fusarium* species infestation of 103% was obtained according to the SHT in the *F. graminearum* subplot due to the contamination with

F. graminearum (99%), and *F. crookwellense* (4%), explaining the high amounts of DON, ZON, and 3-AcDON detected in the whole wheat plant samples at the time of harvest of 107, 6, and 2 mg/kg_{dw}, respectively (**Table 5.1**, **Table S5.6**). In 2011, wheat samples exhibited a somewhat varying concentration pattern throughout the investigation period. Most of the trichothecenes (type A&B) were produced in quantifiable concentrations already three weeks after infection. The concentrations peaked for several of them two to three weeks before harvest (HT-2, T-2, 3-AcDON, DON and NIV; **Table 5.1**), but declined thereafter. This phenomenon was reported before[28-30] and was attributed to the breakdown or conjugation, as well as wash-off of these rather water soluble compounds.[29,31-33] In contrast, BEA and ZON exhibited highest concentrations not until harvest, of 0.7 and 7.4 mg/kg_{dw}, respectively. In both years, type B trichothecenes (DON, NIV, 3-AcDON) and ZON were the most prevalent mycotoxins detected in the subplot infected with *F. graminearum.* This is in line with former studies reviewed by Bottalico et al.[34] Even the co-occurrence of type A trichothecenes with *F. graminearum* infestation was reported before.[35] Generally, *F. graminearum* is one of the most frequently occurring species responsible for *Fusarium* head blight in cereals of Central Europe,[36] and a major producer of DON in wheat grains.[37] The overall lower concentrations detected in 2011 can be attributed to the lower *Fusarium* species infection rate of 89%, compared to 103% in 2011 (**Table S5.6**).

Table 5.1. Incidence of wheat grains infected with *Fusarium* spp. [%], crop yield [kg_{dw}/ha], and mycotoxin concentrations [mg/kg_{dw}] detected in whole wheat plants in all subfields from the upper test field monitored at harvest in 2010 and several times until harvest 2011[a].

Subplot artificially infected with:	Inc. Fus[c] [%]	Yield [t/ha]		Type A trichothecenes: HT-2 [mg/kg_{dw}]		T-2 [mg/kg_{dw}]		Type B trichothecenes: 3-AcDON [mg/kg_{dw}]		DON [mg/kg_{dw}]		FUS-X [mg/kg_{dw}]		NIV [mg/kg_{dw}]		BEA [mg/kg_{dw}]		ZON [mg/kg_{dw}]	
Control	68/-	4.8	*29.07.2010*	*nd*		*nd*		*0.3*		*4.4*		*nd*		*nd*		*0.05*		*0.1*	
			16.06.2011	0.2		0.04		nd		nd		nd		nd		0.07		nd	
			30.06.2011	0.4		0.10		0.08		0.7		nd		0.3		0.05		nd	
	5/-	4.6	14.07.2011	*0.03*	±0.03[b]	*0.02*	±0.06[b]	*0.06*	±0.01[b]	*0.7*	±0.2[b]	*nd*		*0.2*	±0.04[b]	*0.5*	±0.01[b]	*0.03*	±0.04[b]
F. avenaceum	77/73	4.9	*29.07.2010*	*nd*		*nd*		*0.1*		*1.8*		*nd*		*0.2*		*0.02*		*0.2*	
			16.06.2011	0.3		0.05		nd		nd		nd		nd		nd		nd	
			30.06.2011	0.4		0.12		0.2		1.9		nd		0.5		0.02		nd	
	93/92	3.9	14.07.2011	*nd*		*nd*		*nd*		*0.1*	±0.01[b]	*nd*		*nd*		*0.06*	±0.02[b]	*nd*	
F. crookwellense	103/99	3.9	*29.07.2010*	*nd*		*nd*		*0.2*		*3.2*		***nd***		***9.8***		*0.03*		***1.8***	
			16.06.2011	0.5		0.03		nd		-		**nd**		**6.5**		nd		**nd**	
			30.06.2011	0.3		0.1		nd		0.2		**0.7**		**11.9**		-		**0.09**	
	88/86	3.8	14.07.2011	*0.2*	±0.03*	*0.07*	±0.03*	*0.11*	±0.01[b]	*2.1*	±0.08[b]	***0.2***	±0.01[b]	***3.9***	±0.8[b]	*0.7*	±0.2[b]	***0.5***	±0.07[b]
F. graminearum	103/99	1.9	*29.07.2010*	*nd*		*nd*		***1.8***		***107.0***		***nd***		***0.7***		*nd*		***5.7***	
			16.06.2011	0.5		0.1		**0.2**		**30.6**		**nd**		**0.3**		nd		**nd**	
			30.06.2011	0.2		0.1		**3.1**		**133.0**		**nd**		**0.4**		0.4		**0.7**	
	89/88	3.9	14.07.2011	*0.1*	±0.01[b]	*0.03*	±0.02[b]	***2.1***	±0.1[b]	***37.4***	±3[b]	***nd***		***0.9***	±0.1[b]	*0.6*	±0.03[b]	***6.3***	±0.2[b]
F. poae	28/13	4.9	*29.07.2010*	***nd***		***nd***		*nd*		*1.1*		***nd***		***0.9***		***0.4***		*0.08*	
			16.06.2011	**0.4**		**0.1**		nd		nd		**nd**		**nd**		**0.05**		nd	
			30.06.2011	**0.1**		**0.1**		nd		0.2		**nd**		**1.1**		**2.2**		0.04	
	18/12	3.8	14.07.2011	***0.1***	±0.01[b]	***0.03***	±0.02[b]	*nd*		*0.2*	±0.3[b]	***nd***		***0.4***	±0.05[b]	***0.4***	±0.01[b]	*0.3*	±0.03[b]

[a] nd: not detected; Bold numbers indicate mycotoxins associated with the individual *Fusarium* species according to.[28,32] Numbers in italics indicate the time of harvest (in 2011 the samples were collected two days prior to the harvest). [b] Triplicate measurements ± standard deviation of three measurements. [c] The 1st number indicates the percentage of wheat grains infected with *Fusarium* spp and the 2nd number the percentage of grains infected with the corresponding *Fusarium* species. A percentage of >100 is possible if two or more species are isolated from a single grain. In Table S6 for each *Fusarium* species the corresponding incidence is given in detail.

In the *F. crookwellense* subplot, the SHT revealed a total *Fusarium* species incidence in the grains of 103 and 88% in 2010 and 2011, respectively (**Table 5.1**, **Table S5.6**). Here, NIV showed the highest concentration of 3.9–11.9 mg/kg$_{dw}$, both in comparison with the other mycotoxins, and with the NIV content in other subplots, followed by DON and ZON (**Table 5.1**). BEA, HT-2, T-2, FUS-X and 3-AcDON were quantified as well, but in much lower concentrations (0.04–1.8 mg/kg$_{dw}$). Overall, the here presented mycotoxin pattern is in good agreement with the mycotoxins commonly associated with *F. crookwellense.*[34, 38] The co-occurrence of NIV and FUS-X, with DON was particularly reported for small grain cereals in Central and Eastern Europe.[34]

All compounds detected in the wheat samples from the *F. avenaceum* subplot exhibited low concentrations compared to the other subplots, even though a *Fusarium* species incidence of 77 and 93% in 2010 and 2011, respectively, was observed (**Table 5.1**, **Table S5.6**). Mycotoxins commonly associated with *F. avenaceum* are moniliformin and enniatins.[34] Due to analytical difficulties particularly in water samples, and the non-availability of analytical standards, moniliformin and enniatins were not integrated in the analytical method procedure. BEA was detected in very low concentrations of 0.03 and 0.06 mg/kg$_{dw}$ in 2010 and 2011, respectively, and thus can be attributed to the adjacent *F. poae* subplot. Further mycotoxins detected in the wheat plant samples from this subplot probably originated from cross-contamination of wind dispersed conidia of neighboring plots infected with *F. poae*.

Overall, the lowest mycotoxin concentrations were detected in the *F. poae* subplot, which could be explained by the low total incidence documented for this subplot in 2010 and 2011 of 28, and 18%, respectively (**Table 5.1**, **Table S5.6**). Apart from DON and ZON, all other mycotoxins detected in the wheat samples of this subplot are commonly associated with *F. poae.*[39]

The control subplot showed a diverse mycotoxin pattern within the two years of the study. Even though it was not infected by a specific conidia suspension, straw contaminated with *F. graminearum* from the previous year (2009) was placed on the subplot as well. Additionally, the conidia from adjacent subplots could have been easily relocated by the air resulting in fungal infection and mycotoxin production in this subplot. In 2010, four mycotoxins were detected in the wheat plants of the control subplot: 3-AcDON, DON, BEA, and ZON. For DON, the highest concentration of 4.4 mg/kg_{dw} was determined. The three other compounds 3-AcDON, BEA, and ZON were detected with 0.3, 0.05, and 0.2 mg/kg_{dw}, respectively. In 2011, HT-2 and T-2, as well as FUS-X were detected over the three different sampling times within the growing season (**Table 5.1**). Overall, a similar mycotoxin pattern compared to the adjacent subplot infected with *F. graminearum* conidia spore suspension was observed, but at much lower concentrations (0.02–0.7 mg/kg_{dw}).

Based on the detected concentrations of (free, i.e. non-conjugated) mycotoxins in the whole wheat plants at the time of harvest, and the crop yield obtained in each subplot of 1.9–4.9 t_{dw}/ha (Table 2), the individual median (free) mycotoxin production by *Fusarium* species from the experimental winter wheat field site summed up to 292 g/ha in 2010 and 160 g/ha in 2011 in the case of DON (for other mycotoxins, see **Table S5.8**). These numbers are roughly a factor of three to six larger than the 50 g/ha of DON quantified in *F. graminearum* infected wheat plants in an earlier study.[9] For ZON, we calculated amounts of 24 and 28 g/ha in 2010 and 2011, respectively, which is comparable to data from the same site obtained by Hartmann et al.[8] in an extended field study from 2005 to 2007 (6–25 g/ha).

Although the lower field (**Figure S5.1**, II) was investigated within a much smaller period of time, and managed in a different way (see above and **SI**), it largely confirmed the results presented above. Briefly, BEA, ZON, NIV, and DON were detected in the wheat plant samples with concentrations ranging from 0.05 to 1.3 mg/kg_{dw}. The overall crop yield obtained in the lower field was 4.7 t_{dw}/ha (**Table S5.7**), and 0.6, 0.8, 4.2 and 6.1 g mycotoxin/ha for BEA, ZON,

NIV, and DON, respectively, were associated with the wheat plants at the time of harvest in 2011 (for details, see SI-3 and Table S8). Interestingly, even though no additional artificial infection was conducted in the lower field, a *F. poae* incidence of 14%, and a *F. graminearum* incidence of 6%, respectively, was observed (**Table S5.6**). This infection was not solely caused by cross-contamination by wind dispersed conidia of neighboring plots, because the *F. poae* infestation in the upper field showed a lower *F. poae* incidence.

5.3.2 Mycotoxins in drainage water samples

Drainage water was sampled from December 2009 until the end of October 2011. The discharge is depicted in Figure 1A. Note that there was no drainage water discharge during periods of no or only little precipitation (**Figure S5.2**). A water mass balance of both field plots over the investigated time period including precipitation, evapotranspiration, and drainage water discharge revealed a drainage efficiency of 49% for the upper plot, and 21% for the lower plot area (**Table S5.2**). In a previous study under a corn-wheat crop rotation, these numbers were 73 and 44%, respectively.[8] For the lower plot a drainage efficiency of 28% was determined under a grassland-red clover mixture.[40] Overall, the average yearly rainfalls in 2010 and 2011 were considerably lower compared to the previous years,[8,40] which may – together with the different crops – explain the overall lower percentages.

Table 5.2. Mycotoxin concentrations in drainage water samples of *Fusarium* species infected winter wheat crops in [ng/L] collected from December 2009 until the end of October 2011.

Compound	no. of samples analyzed/ above limit of quantification (% of detection)	min concentration [ng/L]	max concentration [ng/L]	average concentration (STD) [ng/L]*
3-AcDON	411/79 (19%)	5.5	367.5	46 (59)
DON	411/222 (54%)	0.8	1114.5	75 (184)
NIV	411/179 (44%)	5.0	71.3	13 (14)
BEA	411/36 (9%)	1.4	10.4	2.3 (2)
ZON	411/14 (3%)	6.0	48.4	12.6 (14)

* Average concentrations calculated solely from the analytes, which were quantifiable. STD = standard deviation calculated from all quantified analytes.

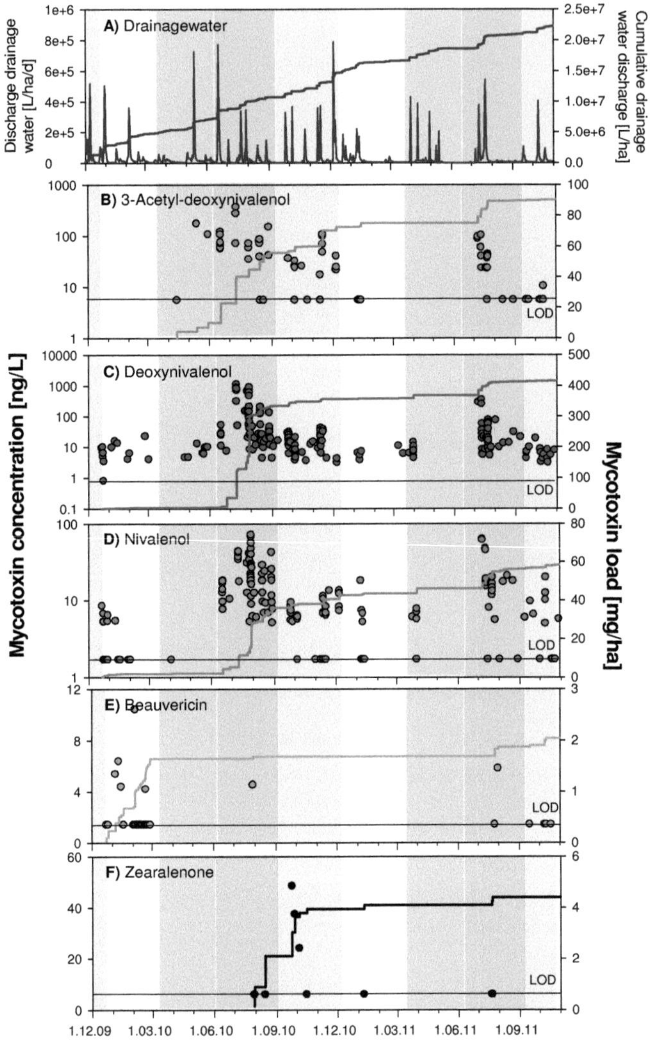

Figure 5.1. Occurrence of mycotoxins in drainage water samples from December 2009 until October 2011 of the upper study field at Agroscope ART: (A) drainage water discharge [L/ha/d] (peaks: discharge drainage water; line: cumulative drainage water discharge); (B) 3-acetyl-deoxynivalenol [ng/L]; (C) deoxynivalenol [ng/L]; (D) nivalenol [ng/L]; (E) beauvericin [ng/L]; (F) zearalenone [ng/L]. [B] to [F]: circles: mycotoxin concentration; line: mycotoxin load. The colored stripes indicate the seasons: spring: green; summer: red; autumn: yellow; winter: white. LOD: limit of detection; points at LOD-level: not detected. Note that panel B, C, and D have primary y-axes in logarithmic scales.

Five mycotoxins (3-AcDON, DON, NIV, ZON, and BEA) out of 33 integrated in the analytical method were regularly quantified in the drainage water samples of the upper field (**Table 5.2**). Although present in wheat, HT-2, T-2, and FUS-X were never detected in the drainage water samples. Generally, concentrations obtained for both fields (I&II) (**Figure S5.1**) were basically similar, although drainage water loads were lower at the lower plot (Figure S4) because of reduced water flow, and the shorter investigation period. DON was detected at highest frequency (54%) and with a mean concentration of 14.9 ng/L. In terms of frequency of occurrence, DON was closely followed by NIV with 44% (mean concentration: 7.7 ng/L), and 3-AcDON with 19% (mean concentration: 25.2 ng/L). BEA and ZON were by far less frequently detected with 9 and 3% (mean concentrations: 1.4, and 6.0 ng/L, respectively). Maximum concentrations of DON, 3-AcDON, NIV, ZON, and BEA were as high as 1.1 µg/L, 367.5, 71.3, 48.4, and 10.4 ng/L, respectively (**Figure 5.1**B-F, **Table 5.2**). Comparable numbers were obtained for DON (23 ng/L – 4.9 µg/L), and ZON (not detected – 35.0 ng/L) in the respective earlier studies.[8,9] Concentrations in the order of a few µg/L for DON were recently also observed in runoff water from *Fusarium* infected wheat heads in greenhouse pot experiments.[41]

The temporal occurrence of the five detected mycotoxins within the 23 months of investigation in drainage water of the upper field is depicted in **Figure 5.1**B-F. Whereas the type B trichothecenes 3-AcDON, DON, and NIV (**Figure 5.1**B, C, and D, respectively) prevailed throughout the year, although with markedly elevated concentrations in summer, ZON concentrations (**Figure 5.1**E) peaked in autumn, and BEA (**Figure 5.1**F) was found predominantly in the winter of 2010 (but not so in 2011, probably because of lower infection rates and mycotoxin production in the preceding wheat cultivation season). The prevalence of mycotoxins in summer seems plausible given the time of *Fusarium* infestation and mycotoxin production in growing wheat plants (see above). Also, the continuous emission of some mycotoxins after harvest and even until spring of the following year can be rationalized with ongoing elution of mycotoxins from wheat straw left on the field after harvest.

After nearly two years of investigation, the total (free) mycotoxin loads exported via drainage water were 416, 90, 58, 4.4, and 2.1 mg/ha for DON, 3-AcDON, NIV, ZON, and BEA, respectively (**Figure 5.1**B-F, right y-axis). In general, these loads were slightly underestimated because the hydrological capacity of the drainage water sampling device was limited to 2.5 L/s. This range was exceeded at several occasions, during which no samples proportional to the flow could be taken. Regarding the seasonal emission dynamics, the main precipitation occurred in summer (36% of the total precipitation) generating 91% of the DON load. In contrast, during winter only 16% of the total precipitation was observed, and the DON load emitted in this period corresponded to just 1%. The same seasonal emission dynamics were observed for 3-AcDON and NIV, whereas during autumn 21% of the total precipitation caused 66% of the ZON load. Interestingly, for BEA, the 21% of precipitation in autumn caused just 9% of the detected loads, whereas only 16% of the total precipitation during winter led to main emission loads of 75% for BEA. Together, the amounts produced *in planta* by the different *Fusarium* species at the time of harvest in 2010 and 2011 (see above) and the mycotoxin loads detected in the respective season translate into emitted fractions as follows: DON: 0.03/0.12%, NIV: 0.07/0.09%, 3-AcDON: 0.01/0.11%, ZON: 0.002/0.016%, BEA: 0.005/0.066%. These numbers differ at least by a factor of three from those reported earlier for DON (1.2%[9]), and for ZON (0.07%[8,9]).

In the lower field, a similar emission dynamic was observed as reported for the upper subfield, except that 3-AcDON was not detected in any of the collected drainage water samples. Over the five month of investigation, peak concentrations for DON and NIV were detected in July before and after harvest with 103 and 59 ng/L, respectively. BEA and ZON reached peak concentrations in September 2011 of 6.5, and 12.7 ng/L, respectively. The total mycotoxin loads exported via drainage water were 15, 19, 0.8, and 0.6 mg/ha for DON, NIV, ZON, and BEA, respectively (**Table S5.6**). These loads correspond to about 0.24,

0.45, 0.12 and 0.10% for DON, NIV, ZON, and BEA, respectively, of the total amounts of mycotoxins present in the whole wheat plants at the time of harvest.

Qualitatively, both the seasonal occurrence and the emitted fractions of mycotoxins in drainage water can be explained with their physical-chemical properties, in particular, their hydrophobicity as expressed by the soil organic matter-water distribution coefficient (Log K_{oc}), or if unavailable the octanol-water partitioning coefficient (Log K_{ow}; Table 1). With Log K_{oc} numbers below 0.7 ($0 \leq$ estimated Log $K_{ow} \leq 1.7$), the three type B trichothecenes DON, NIV and 3-AcDON are by far the most hydrophilic, and thus most mobile, of all detected compounds. Hence, they are expected to be least retarded in the soil passage, and to elute the most. Conversely, ZON (Log K_{oc}=3.42, Log K_{ow}=3.6; **Table 5.3**) and BEA (Log K_{oc}=5.9, Log K_{ow}=5.5; **Table 5.3**) are considerably more hydrophobic, and should thus be substantially retained in soils. Related to this fact, first flushes observed in this study were more pronounced for hydrophilic (DON; NIV, and 3-AcDON) than hydrophobic compounds (ZON and BEA), although they may also take place for the some of the latter ones[42] (**Figure S5.5**).

Table 5.3. Physical-chemical properties of the mycotoxins regularly quantified in wheat and/or drainage water.

	Type A trichothecenes		Type B trichothecenes					
	HT-2	T-2	3-AcDON	DON	FUS-X	NIV	BEA	ZON
Molecular structure	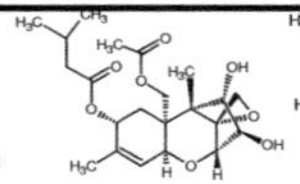	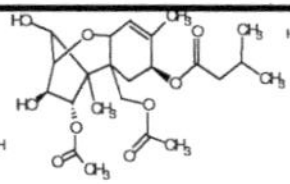	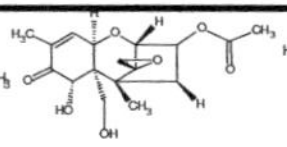	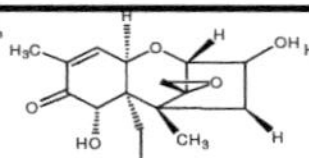	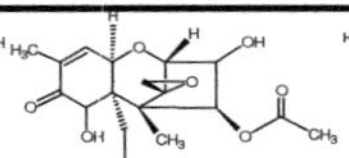	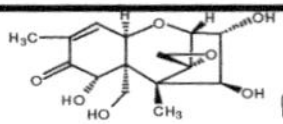	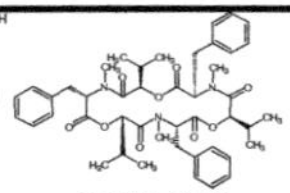	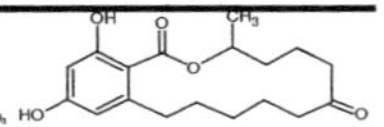
CAS number	26934-87-2	36519-25-2	50722-38-8	51481-10-8	23255-69-8	23282-20-4	26048-05-5	17924-92-4
Molecular formula	$C_{22}H_{32}O_8$	$C_{19}H_{26}O_8$	$C_{17}H_{22}O_7$	$C_{15}H_{20}O_6$	$C_{17}H_{22}O_8$	$C_{15}H_{20}O_7$	$C_{45}H_{57}N_3O_9$	$C_{18}H_{22}O_5$
Molecular weight [g/mol]	424.5	382.4	338.4	296.3	354.4	312.3	783.9	318.4
Log K_{ow}	$2.2^{a,b}$	$2.8^{a,b}$	$1.3^{a,b}$	$0.1^{a,b}$	$1.4^{a,b}$	$0^{a,b}$	5.5^{c}	$3.6^{a,b}$
Log K_{oc}	1.01^{b}	1.03^{b}	$< 0.7^{b}$	$< 0.7^{b}$	$< 0.7^{b}$	$< 0.7^{b}$	$5.9^{a,b}$	3.42^{b}
pK_a	13.3^{d}	13.2^{d}	11.8^{d}	11.9^{d}	11.7^{d}	11.8^{d}	-	7.6^{d}
Half-life in natural waters [d]^{c}	60	60	55	38	38	38	180	38
Produced by *Fusarium* species	*F. poae*	*F. poae*	*F. graminearum*	*F. graminearum*	*F. poae, F. graminearum, F. crookwellense*	*F. poae, F. graminearum, F. crookwellense*	*F. avenaceum, F. poae*	*F. graminearum F. crookwellense*

^{a} Calculation with COSMOtherm. ^{b} From ref[20]. ^{c} Calculation with EPISuite v4.0. ^{d} Calculation with SPARC v4.5. (-, No data available)

5.3.3 Environmental and ecotoxicological relevance

The results of this extended study underline and confirm earlier data that surface waters are – in absolute terms – just a minor recipient of mycotoxins from *Fusarium* spp. infected arable land.[8-10] However, it should be emphasized that the mycotoxin amounts produced in the current study are similar to those of agricultural application rates of many modern pesticides (50 g – 1 kg/ha per whole season). Additionally, the emitted concentrations of, e.g. DON from drainage water (max. of 1.1 µg/L) were similar to those of, e.g., atrazine in small agricultural catchments.[43] Moreover, the emitted fractions are as well comparable to those reported for more notorious agro-environmental chemicals such as atrazine or veterinary sulfonamide antibiotics.[43,44]

For all compounds detected in the drainage water, the large dilution in surface waters led to the drop of the quantified concentrations to even below the detection limits,[9,45] and ultimately most probably to below critical environmental levels. However, ZON for instance might contribute to an overall estrogenic contamination in small creeks which are mainly fed by agricultural run-off,[8] and caused effects in life-cycle exposure experiments with zebra fish in the sub µg/L concentration range.[46] As for the other mycotoxins, especially the type B trichothecenes, their ecotoxicity is still largely unknown. Recent investigations show that NIV is substantially more toxic both towards mammalians[47] and insects (lepidopteran *Spodoptera frugiperda* (SF-9) insect cell line[48]), although it only differs from DON by the presence of a hydroxyl group at C4. The acute toxicity of BEA on freeze-dried bioluminescent bacteria *Vibrio fisheri* was measured by bacterial bioluminescence assay and reported to be moderate with an EC_{50} of 94±9 µg/ml.[49]

As a consequence of the here presented study, we consider it advisable to investigate the environmental exposure of natural toxins, such as mycotoxins, phytoestrogens and various plant toxins,[50] in further detail, because these compounds add to the mixture of many already known aquatic micropollutants,[51] while their ecotoxicological risk is by far less evaluated.

Acknowledgment

We thank the Swiss Federal Office for the Environment for the financial support. A special thank goes to the field staff at Agroscope ART and their great support. Additionally, we would like to commemorate Andreas Hecker, who was always very helpful in setting up the field site, inoculating the plants and had a profound knowledge on *Fusarium* diseases. Finally, we would like to acknowledge Pakeerathan Srikanthan and Stefan Weissen for their excellent help with the SPE extractions.

References (chapter 5)

1. Desjardin, A. E., *Fusarium Mycotoxins - Chemistry, Genetics and Biology*. 1. ed.; The American Phytopathological Society: St. Paul, Minnesote 55121, U.S.A., 2006; Vol. 1, p 240.
2. Bennett, J. W.; Klich, M., Mycotoxins. *Clin. Microbiol. Rev.* **2003,** *16*, 497-516.
3. Hussein, H. S.; Brasel, J. M., Toxicity, metabolism, and impact of mycotoxins on humans and animals. *Toxicology* **2001,** *167*, 101-134.
4. Kuiper-Goodman, T., Food safety: Mycotoxins and phycotoxins in perspective. In *Mycotoxins and Phycotoxins - Developments in Chemistry, Toxicology and Food Safety*, Miraglia, M.; VanEgmond, H. P.; Brera, C.; Gilbert, J., Eds. Alaken, Inc: Ft Collins, 1998; pp 25-48.
5. Krska, R.; Schubert-Ullrich, P.; Molinelli, A.; Sulyok, M.; Macdonald, S.; Crews, C. In *Mycotoxin analysis: An update*, 12th IUPAC International Symposium on Mycotoxins and Phycotoxins, Istanbul, TURKEY, May 21-25, 2007; Taylor & Francis Ltd: Istanbul, TURKEY, 2007; pp 152-163.
6. Shephard, G. S., Determination of mycotoxins in human foods. *Chem. Soc. Rev.* **2008,** *37*, 2468-2477.
7. Hartmann, N.; Erbs, M.; Wettstein, F. E.; Schwarzenbach, R. P.; Bucheli, T. D., Environmental exposure to estrogenic and other myco- & phytotoxins. *Chimia* **2008,** *62*, 364-367.
8. Hartmann, N.; Erbs, M.; Wettstein, F.E.; Schwarzenbach, R.P.; Bucheli, T. D., Occurrence of zearalenone on *Fusarium graminearum* infected wheat and maize

fields in crop organs, soil and drainage water. *Environ. Sci. Technol.* **2008,** *42*, 5455-5460.

9. Bucheli, T. D.; Wettstein, F. E.; Hartmann, N.; Erbs, M.; Vogelgsang, S.; Forrer, H.-R.; Schwarzenbach, R. P., Fusarium mycotoxins: Overlooked aquatic micropollutants? *J. Agric. Food Chem.* **2008,** *56*, 1029-1034.
10. Kolpin, D. W.; Hoerger, C. C.; Meyer, M. T.; Wettstein, F. E.; Hubbard, L. E.; Bucheli, T. D., Phytoestrogens and mycotoxins in Iowa Streams: An examination of underinvestigated compounds in agricultural basins. *J. Environ. Qual.* **2010,** *39*, 2089-2099.
11. Schenzel, J.; Schwarzenbach, R. P.; Bucheli, T. D., A multi residue screening method to quantify mycotoxins in aqueous environmental samples. *J. Agric. Food Chem.* **2010,** *58*, 11207-11217.
12. Bester, K.; McArdell, C. S.; Wahlberg, C.; Bucheli, T. D., Quantitative mass flows of selected xenobiotics in urban waters and waste water treatment plants. In *Enviornmental Pollution* Fatta-Kassinos, D.; Bester, K.; Kuemmerer, K., Eds. Springer Netherlands: Dordrecht, 2010; Vol. 16.
13. Wettstein, F. E.; Bucheli, T. D., Poor elimination rates in waste water treatment plants lead to continuous emission of deoxynivalenol into the aquatic environment. *Wat. Res.* **2010,** *44*, 4137-4142.
14. FAO, *World reference base for soil resources*. 84 ed.; International Soil Reference and Information Centre: Viale delle Terme di Caracalla, 100 Rome, Italy., 1998.
15. Champeil, A.; Dore, T.; Fourbet, J. F., Fusarium head blight: epidemiological origin of the effects of cultural practices on head blight attacks and the production of mycotoxins by Fusarium in wheat grains. *Plant Science* **2004,** *166*, 1389-1415.
16. Blandino, M.; Reyneri, A.; Vanara, F.; Tamietti, G.; Pietri, A., Influence of agricultural practices on *Fusarium* infection, fumonisin and deoxynivalenol contamination of maize kernels. *World Mycotoxin J.* **2009,** *2*, 409-418.
17. Champeil, A.; Fourbet, J. F.; Dore, T.; Rossignol, L., Influence of cropping system on Fusarium head blight and mycotoxin levels in winter wheat. *Crop Protection* **2004,** *23*, 531-537.

18. Doohan, F. M.; Brennan, J.; Cooke, B. M., Influence of climatic factors on *Fusarium* species pathogenic to cereals. *Eur. J. Plant Pathol.* **2003,** *109*, 755-768.

19. Malihipour, A.; Gilbert, J.; Piercey-Normore, M.; Cloutier, S., Molecular phylogenetic analysis, trichothecene chemotype patterns, and variation in aggressiveness of *Fusarium* isolates causing head blight in wheat. *Plant Disease* **2012,** *96*, 1016-1025.

20. Parry, D. W.; Jenkinson, P.; McLeod, L., *Fusarium* ear blight (scab) in small-grain cereals - A review. *Plant Pathol.* **1995,** *44*, 207-238.

21. Vogelgsang, S.; Sulyok, M.; Hecker, A.; Jenny, E.; Krska, R.; Schuhmacher, R.; Forrer, H. R., Toxigenicity and pathogenicity of *Fusarium poae* and *Fusarium avenaceum* on wheat. *Eur. J. Plant Pathol.* **2008,** *122*, 265-276.

22. Leslie, J. F.; Summerell, B. A., *The Fusarium laboratory manual.* Iowa, U.S.A., 2006; p 388.

23. Schenzel, J.; Forrer, H.-R.; Vogelgsang, S.; Bucheli, T., Development, validation and application of a multi-mycotoxin method for the analysis of whole wheat plants. *Mycotoxin Research* **2012,** *28*, 1-13.

24. Scudamore, K. A.; Livesey, C. T., Occurrence and significance of mycotoxins in forage crops and silage: a review. *J. Sci. Food Agric.* **1998,** *77*, 1-17.

25. Muller, M. E. H.; Koszinski, S.; Brenning, A.; Verch, G.; Korn, U.; Sommer, M., Within-field variation of mycotoxin contamination of winter wheat is related to indicators of soil moisture. *Plant and Soil* **2011,** *342*, 289-300.

26. Vogelgsang, S.; Hecker, A.; Musa, T.; Dorn, B.; Forrer, H.-R., On-farm experiments over 5 years in a grain maize/winter wheat rotation: effect of maize residue treatments on *Fusarium graminearum* infection and deoxynivalenol contamination in wheat. *Mycotoxin Research* **2011,** *27*, 81-96.

27. Chelkowski, J.; Gromadzka, K.; Stepien, L.; Lenc, L.; Kostecki, M.; Berthiller, F., *Fusarium* species, zearalenone and deoxynivalenol content in preharvest scabby wheat heads from Poland. *World Mycotoxin J.* **2012,** *5*, 133-141.

28. Matthaus, K.; Danicke, S.; Vahjen, W.; Simon, O.; Wang, J.; Valenta, H.; Meyer, K.; Strumpf, A.; Ziesenib, H.; Flachowsky, G., Progression of mycotoxin and nutrient concentrations in wheat after inoculation with Fusarium culmorum. *Arch. Ani. Nutri.* **2004,** *58*, 19-35.

29. Rohweder, D.; Valenta, H.; Sondermann, S.; Schollenberger, M.; Drochner, W.; Pahlow, G.; Doll, S.; Danicke, S., Effect of different storage conditions on the mycotoxin contamination of *Fusarium culmorum*-infected and non-infected wheat straw. *Mycotoxin Research* **2011,** *27*, 145-153.

30. Brinkmeyer, U.; Danicke, S.; Lehmann, M.; Valenta, H.; Lebzien, P.; Schollenberger, M.; Sudekum, K. H.; Weinert, J.; Flachowsky, G., Influence of a *Fusarium culmorum* inoculation of wheat on the progression of mycotoxin accumulation, ingredient concentrations and ruminal in sacco dry matter degradation of wheat residues. *Arch. Ani. Nutri.* **2006,** *60*, 141-157.

31. Engelhardt, G.; Ruhland, M.; Wallnofer, P. R., Metabolism of mycotoxins in plants. *Advances in Food Sciences* **1999,** *21*, 71-78.

32. Berthiller, F.; Dall'Asta, C.; Schuhmacher, R.; Lemmens, M.; Adam, G.; Krska, R., Masked mycotoxins: Determination of a deoxynivalenol glucoside in artificially and naturally contaminated wheat by liquid chromatography-tandem mass spectrometry. *J. Agric. Food Chem.* **2005,** *53*, 3421-3425.

33. Nakagawa, H.; Ohmichi, K.; Sakamoto, S.; Sago, Y.; Kushiro, M.; Nagashima, H.; Yoshida, M.; Nakajima, T., Detection of a new *Fusarium* masked mycotoxin in wheat grain by high-resolution LC-Orbitrap (TM) MS. *Food Addit. Contam. Part A-Chem.* **2011,** *28*, 1447-1456.

34. Bottalico, A.; Perrone, G., Toxigenic *Fusarium* species and mycotoxins associated with head blight in small-grain cereals in Europe. *Eur. J. Plant Pathol.* **2002,** *108*, 611-624.

35. Muller, H. M.; Schwadorf, K., A Survey of the natural occurrence of *Fusarium* toxins in wheat grown in a Southwestern area of Germany. *Mycopathologia* **1993,** *121*, 115-121.

36. Gale, L. R., Population biology of *Fusarium* species causing head blight of grain crops. In *Fusarium head blight of wheat and barley*, Leonard, K. J., Ed. APS Press, St. Paul: 2003; pp 120-143.

37. Richard, J. L.; Payne, G. A.; Desjardin, A. E.; Maragos, C. M., *Mycotoxins: Risk in Plant, Animal, and Human Systems*. Library of Congress Cataloging in Publication Data: Ames, Iowa, U.S.A., 2003; p 191.

38. Buck, H. T.; Nisi, J. E.; Salomón, N., *Wheat Production in Stressed Environments*. Springer: Dordrecht, The Netherlands, 2007; Vol. 1.

39. Logrieco, A.; Mule, G.; Moretti, A.; Bottalico, A., Toxigenic *Fusarium* species and mycotoxins associated with maize ear rot in Europe. *Eur. J. Plant Pathol.* **2002,** *108*, 597-609.

40. Hoerger, C. C.; Wettstein, F. E.; Bachmann, H. J.; Hungerbuhler, K.; Bucheli, T. D., Occurrence and mass balance of isoflavones on an experimental grassland field. *Environ. Sci. Technol.* **2011,** *45*, 6752-6760.

41. Gautam, P.; Dill-Macky, R., Free water can leach mycotoxins from *Fusarium*-infected wheat heads. *J. Phytopathol.* **2012,** *160*, 484-490.

42. Hartmann, N.; Erbs, M.; Wettstein, F. E.; Schwarzenbach, R. P.; Bucheli, T. D., Quantification of estrogenic mycotoxins at the ng/L level in aqueous environmental samples using deuterated internal standards. *J. Chromatogr. A* **2007,** *1138*, 132-140.

43. Leu, C.; Singer, H.; Stamm, C.; Muller, S. R.; Schwarzenbach, R. P., Simultaneous assessment of sources, processes, and factors influencing herbicide losses to surface waters in a small agricultural catchment. *Environ. Sci. Technol.* **2004,** *38*, 3827-3834.

44. Stoob, K.; Singer, H. P.; Mueller, S. R.; Schwarzenbach, R. P.; Stamm, C. H., Dissipation and transport of veterinary sulfonamide antibiotics after manure application to grassland in a small catchment. *Environ. Sci. Technol.* **2007,** *41*, 7349-7355.

45. Schenzel, J.; Hungerbuhler, K.; Bucheli, T. D., Mycotoxins in the environment: II. Occurrence and origin of mycotoxins in Swiss river waters. *Environ. Sci. Technol.* **submitted.**

46. Schwartz, P.; Bucheli, T. D.; Wettstein, F. E.; Burkhardt-Holm, P., Life-cycle exposure to the estrogenic mycotoxin zearalenone affects zebrafish (*Danio rerio*) development and reproduction. *Environmental Toxicology* **2011**.

47. Ueno, Y., *Trichothecenes - Chemical, Biological and Toxicological Aspects.* Kodansha/Elsevier: Tokio/Amsterdam, 1983; p 313.

48. Fornelli, F.; Minervini, F.; Mule, G., Cytotoxicity induced by nivalenol, deoxynivalenol, and fumonisin B, in the SF-9 insect cell line. *In Vitro Cell. Dev. Biol.-Anim.* **2004,** *40*, 166-171.

49. Fotso, J.; Smith, J. S., Evaluation of beauvericin toxicity with the bacterial bioluminescence assay and the Ames mutagenicity bioassay. *J. Food Sci.* **2003,** *68*, 1938-1941.

50. Hoerger, C. C.; Schenzel, J.; Strobel, B. W.; Bucheli, T. D., Analysis of selected phytotoxins and mycotoxins in environmental samples. *Anal. Bioanal. Chem.* **2009,** *395*, 1261-1289.

51. Schwarzenbach, R. P.; Escher, B. I.; Fenner, K.; Hofstetter, T. B.; Johnson, C. A.; von Gunten, U.; Wehrli, B., The challenge of micropollutants in aquatic systems. *Science* **2006,** *313*, 1072-1077.

52. Schenzel, J.; Goss, K.-U.; Schwarzenbach, R. P.; Bucheli, T. D.; Droge, S. T. J., Experimentally determined soil organic matter-water sorption coefficients for different classes of natural toxins and comparison with estimated numbers. *Environ. Sci. Technol.* **2012**, *46*, 6118-6126.

Supporting Information

SI-5.1 Field site description, instrumentation, cultivation, and information on the water mass balance.

The ground water table was monitored in the first year of the drainage installation in 30 minutes intervals by three piezometers, which were located on a transect on both field plots. The depth of the ground water table ranged from 100 to 125 cm, for details see.[1] Thus, the ground water table was proven to be always below the drainage system.

Table S5.1. Field soil parameters in all soil horizons (horizons classified according to the World reference base for soil resources) [a]

Soil horizon[b]	Sampling depth [cm]	pH ($CaCl_2$)	C_{org} [%]	Humus [c][%]	Clay[%]	Silt [%]	Sand [%]
Ahp	0 - 20	6.8	2.0	3.4	30.6	39.4	30.0
Bg	20 - 35	7.0	2.6	4.5	44.2	28.9	26.9
Bgg	35 - 50	7.0	0.9	1.5	30.0	38.2	31.8
BCgg	50 - 70	7.7	0.2	0.4	18.2	49.6	32.2
Cgg	70 - 90	7.8	0.1	0.1	13.2	66.3	20.5

[a] Data collected at Agroscope Reckenholz-Taenikon ART, [b] [2], [c] Calculated: C_{org} multiplied by a factor of 1.72

Figure S5.1. Field site at Agroscope ART, A) layout of both study sites (I&II), represented by the green areas. The blue lines in A) and B) indicate the tubes of the two the drainage water systems. The two yellow circles indicate the two sampling ducts of the two study sites. B) Schematic overview of the plot alignment for the artificial Fusarium species infestation.

B)

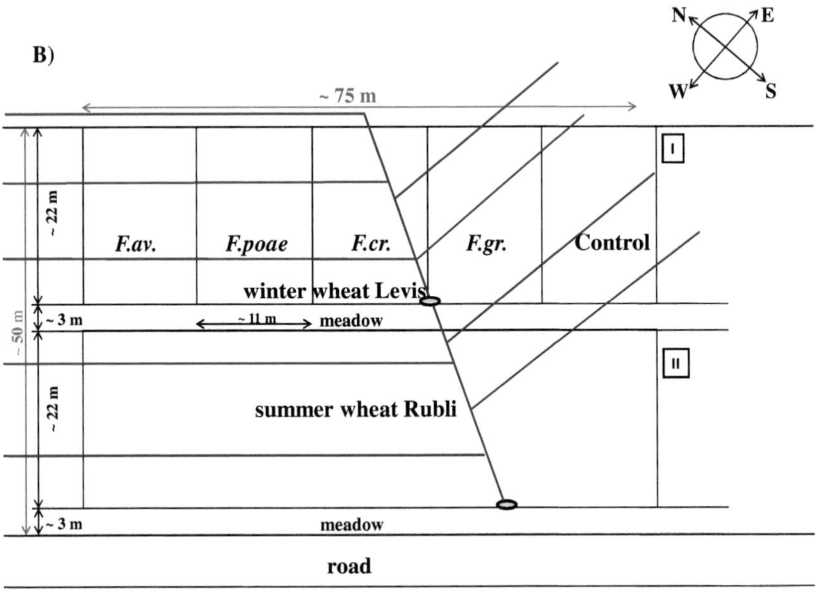

I. Upper field cropped with winter wheat of the variety Levis. Artificial infections: *F.av.*: *Fusarium avenaceum*, *Fusarium poae*, *F.cr.*: *Fusarium crookwellense*, *F.gr.*: *Fusarium graminearum*, Control: subplot without artificial infection.

II. Lower field cropped with summer wheat of the variety Rubli without artificial infection.

Figure S5.2. A) Daily precipitation (gathered at intervals of 10 minutes) at Reckenholz from December 2009 until the end of October 2011 in the upper subplot of the test field. The colored stripes indicate the seasons: spring: green, summer: red, autumn: yellow, winter: white. B) Drainage water discharge [L/(ha*d)].

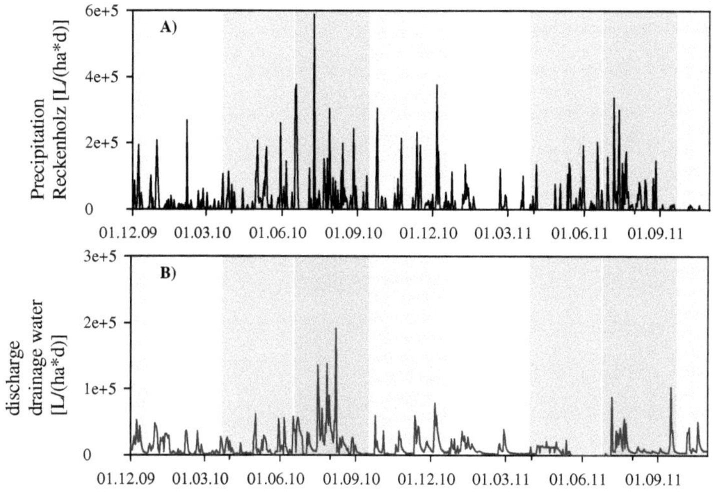

Table S5.2. Water mass balance of the field sites.

	Upper field site [L/0.2 ha][b]	Lower field site [L/0.2 ha][c]
Precipitation	3‘597‘600	853‘600
Evapotranspiration[a]	1‘798‘800	426‘800
Drainage water	889‘032	90‘813
% of the water drained	**49%**	**21%**

[a]Evapotranspiration of plant transpiration multiplied by precipitation. Evapotranspiration of wheat = 50°% (personal communication from P.L. Calanca (ART)); [b] Investigation period: 23 month; [c] Investigation period: 5 month.

The water balance is calculated as follows.[3] drainage water divided by (precipitation minus evapotranspiration) multiplied by 100.

SI-5.2 Analytical procedures and figures of merit. Wheat samples.

All samples were first freeze dried at -20°C for several days and then further dried at 35°C until weight consistency was achieved, but at least for 72 h. Afterwards, the samples were ground and sieved to 2 mm using a SM 2000 cutting mill (Retsch GmbH, Haan, Germany). Samples were processed within 48h after drying. Before extraction, the ground whole wheat plant powder was homogenized with a rotating mixing machine (Turbula® Type T2 F, Willy A. Bachofen AG, Basel, Switzerland) for 15 min. Solid-liquid extraction of 2 g plant material with 20 mL MeCN/Milli-Q water (80:20, v/v) was carried out for 2 h on a SM 30 orbital shaker at 200 rpm (Edmund Bühler GmbH, Hechingen, Germany). After extraction, the suspended material was filtered over a paper filter No. 595½ with a diameter of 110 mm (Schleicher & Schuell GmbH, Dassel, Germany). For clean-up, all of the filtrate was passed through a Varian Bond Elut Mycotoxin® cartridge (VarianInc. Lake Forest CA, USA) and the eluate was collected in 10 mL conical reaction vial vessels (Supelco, Bellfonte PA, USA). Afterwards, 1 mL of each extract was transferred into 5 mL conical reaction vial vessels and the aliquots were reduced to 100 µL under a gentle nitrogen gas stream at 50°C. The extracts were reconstituted in 900 µL of Milli-Q water/MeOH (90:10, v/v) and transferred into 1.5 mL amber glass vials. The samples were stored at +4°C and analyzed within 48 h. Quantification took place with matrix-matched calibrations, produced freshly for each batch of samples. Matrix-matched calibration curves were obtained by carrying out standard addition of the multicomponent stock solution to the final extracts. Curves were obtained by plotting measured analyte peak areas against corresponding analyte concentrations in the extracted matrix. Linear regression was performed for each curve.

Table S5.3. Ion suppressions (IS), absolute recoveries (R), method precisions (MP), and limits of detection and quantification (LOD and LOQ) of mycotoxins in whole wheat plants.

Mycotoxin	50 ng/g		100 ng/g		150 ng/g		LOD[a]	LOQ[a]	IS
	R [%]	MP	R [%]	MP	R [%]	MP	[ng/g]	[ng/g]	[%]
Beauvericin	115	6	110	6	98	10	6	19	-89
Zearalenone	98	21	100	9	96	5	3	11	61
HT-2 toxin	104	6	101	3	98	5	6	20	80
T-2 toxin	104	1	103	2	96	5	5	17	43
3-Acetyl-DON	103	2	103	3	96	3	9	31	53
Deoxynivalenol	98	10	101	2	101	2	3	11	77
Fusarenone-X	102	4	102	2	100	3	19	64	64
Nivalenol	101	4	100	4	90	10	19	63	84

[a,b] Determined using a spiked wheat sample of 50 ng/g.

SI-5.3 Analytical procedures and figures of merit. Aqueous samples.

Raw water samples were filtered (glass fibre filters, pore size 1.2 μm, Millipore, Volketswil, Switzerland) by vacuum filtration (Supelco, Bellfonte PA, USA), transferred to 1 L glass bottles and stored in the dark at +4 °C until analysis within 2 weeks (storage tests showed that mycotoxins were stable over this period of time). Before SPE, the pH was adjusted to between 6.6 and 7.0 by adding either ammonium acetate or acetic acid. In routine analysis, the exact volume of 1 L was spiked with 50 μL of the isotope labeled internal standard (ILIS) mixture before the storage or processing of the sample. The samples were shaken vigorously before further treatment. Filtered water samples (1 L) were concentrated and purified by performing reversed-phase SPE (Oasis HLB cartridges, 6 mL, 200 mg; Waters Corporation, Milford MA, USA) on a 12-fold vacuum extraction box (Supelco, Bellfonte PA, USA). The SPE cartridges were consecutively conditioned with 5 mL of MeOH, 5 mL of Milli-Q water and MeOH (1/1, v/v), and 5 mL of Milli-Q water. One Liter water samples were passed through the cartridges with a maximum flow rate of 10 mL/min. Subsequently, the cartridges were washed with 5 mL of Milli-Q water. Without any additional column drying step, the analytes were eluted with 5 mL MeOH and the aliquots were collected in conical reaction vial vessels (Supelco, Bellfonte PA, USA). The 5 mL MeOH aliquots were reduced to 100 μL under a gentle nitrogen gas stream at 50°C. The extracts were reconstituted in 900 µL of Milli-Q water/MeOH (90/10, v/v) and transferred into 1.5 mL amber glass vials. The samples were stored at +4°C and analyzed within 48 h.

Table S5.4. Ion suppressions (IS), relative recoveries[a], absolute recoveries[a], and method precision (MP) of mycotoxins in drainage water.

Compound	IS [%]	Concentration [ng/L]	Relative Recoveries [%]	Absolute Recoveries [%]	MP [%]
		5	n.a.	n.a.	n.a.
3-Acetyl-DON	-4	25	109 (3)	99 (7)	7
		100	101 (2)	96 (5)	5
		5	n.a.	n.a.	n.a.
Deoxynivalenol	-14	25	99 (4)	99 (4)	4
		100	106 (2)	99 (3)	3
		5	n.d.	59 (10)	17
Nivalenol	20	25	n.d.	58 (3)	5
		100	n.d.	55 (4)	7
		5	n.d.	66 (9)	14
Beauvericin	15	25	n.d.	60 (2)	3
		100	n.d.	84 (10)	12
		5	n.a.	n.a.	n.a.
Zearalenone	12	25	97 (4)	104 (7)	7
		100	102 (1)	95 (4)	4

n.a. not available due to higher method detection limit; [a] Absolute standard deviation (five replicates) in parenthesis; n.d.: not detected/quantifiable due to the absence of isotopic labeled internal standards.

Table S5.5. Limit of detection [ng/L][a] of mycotoxins in drainage water .

Compound	LOD
3-Acetyl-Deoxynivalenol	5.5[b]
Deoxynivalenol	0.8
Nivalenol	1.7
Beauvericin	1.4
Zearalenone	6.0

[a] Three times the absolute standard deviation at 5 ng/L; [b] Three times the absolute standard deviation at 25 ng/L.

SI-5.4 Chromatographic separation and mass spectrometric detection.

LC-MS/MS was performed on a Varian 1200L LC-MS instrument (VarianInc, Walnut Creek CA, USA). The mycotoxins were separated using a 125 mm × 2.0 mm i.d., 3 µm, Pyramid C_{18} column, with a 2.1 mm x 20 mm i.d. 3 µm guard column of the same material (Macherey-Nagel GmbH & Co. KG, Düren, Germany) at room temperature. Two different chromatographic runs were used for the separation of all compounds, one in the positive and one in the negative ionization mode, respectively. The optimized LC mobile phase gradient for the analysis of the analytes measured in negative ionization mode was as follows: 0.0 min: 0% B (100% A), 2.0 min: 0% B, 15.0 min: 100% B, 18.0 min: 100% B, 19.0 min: 0% B, 24.0 min: 0% B. The analytes measured in the positive ionization mode were separated with the following gradient: 0.0 min: 27% B (73% A), 1.0 min: 27% B, 1.3 min: 45% B, 5.0 min: 45% B, 5.3 min: 63% B, 9.0 min: 63% B, 9.3 min: 81% B, 13.0 min: 81% B, 13.3 min: 100% B, 20.0 min: 100% B, 21.0 min: 27% B, 24.0 min: 27% B. In both cases eluent A consisted of Milli-Q water/MeOH/acetic acid (89/10/1, v/v/v) and eluent B of Milli-Q water/MeOH/acetic acid (2/97/1, v/v/v). Both eluents were buffered with 5 mM ammonium acetate. The injection volume was 40 μL and the mobile phase flow was 0.2 mL/min.

LC-MS interface conditions for the ionization of the acidic mycotoxins in the –ESI mode were: needle voltage - 4000 V, nebulizing gas (compressed air) 3.01 bar, drying gas (N_2, 99.5%) 275°C and 1.24 bar, shield voltage -600 V. The neutral mycotoxins were ionized in the +ESI mode: needle voltage + 4500 V, nebulizing gas (compressed air) 3.01 bar, drying gas (N_2, 99.5%) 275°C and 1.24 bar, shield voltage + 600 V.

SI-5.5 Comprised seed health test data for all subfields and for both years of the investigation period 2010, and 2011.

Table S5.6. Incidence of *Fusarium* species on wheat grains as obtained by a seed health test in 2010 and 2011.

Subfield	Year	% grains with infestation **Fg**	**Fav**	**Fp**	**Fcr**	**Fspp**	total infested grains [%]
Control plot	2010	58	3	2	4	1	68
	2011	2	1	1	0	1	5
F. avenaceum	2010	2	73	0	2	0	77
	2011	0	92	1	0	0	93
F. crookwellense	2010	5	0	0	99	0	103
	2011	1	0	1	86	1	88
F. graminearum	2010	99	0	0	4	0	103
	2011	88	0	0	1	0	89
F. poae	2010	8	3	13	3	2	28
	2011	2	4	12	1	0	18
summer wheat Rubli	2011	6	1	14	0	1	22

Fg: *F. graminearum*, Fav: *F. avenaceum*, Fp: *F. poae*, Fcr: *F. crookwellense*, Fspp: *Fusarium* species.

SI-5.6 Complete data from the lower plot during the investigation period from June to October 2011.

Figure S5.3. Daily precipitation (gathered at intervals of 10 minutes) at Reckenholz from June 2011 until mid of October 2011 in the lower subplot of the test field.

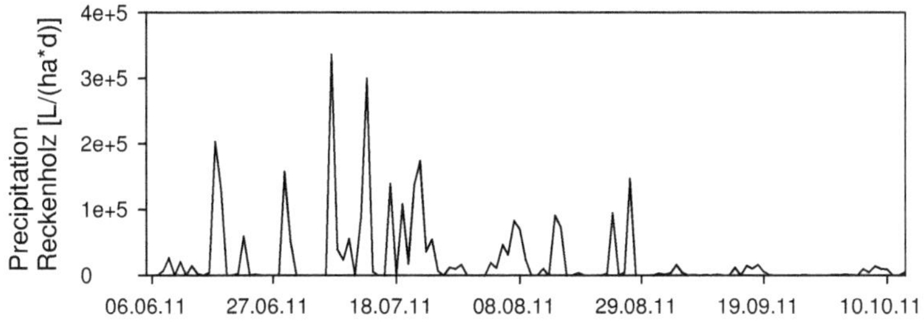

Table S5.7. Concentrations [mg/kg$_{dw}$] of the mycotoxins detected in whole wheat plants from the lower field cropped in 2011 with spring wheat of the variety Rubli monitored several times during the growing season. The yield obtained harvest, emitted mycotoxin load via drainage [mg/ha], and the emitted mycotoxin fraction [%]. Numbers written in italics indicate the time of harvest.

	Yield		Trichothecenes B							
Lower subplot	[t/ha]	[mg/kg$_{dw}$]	DON		NIV		BEA		ZON	
Summer wheat		16.06.2011	nd		0.4		0.05		nd	
		30.06.2011	0.3		0.4		0.14		nd	
		14.07.2011	0.7	±0.06*	0.9	±0.07*	0.13	±0.04*	0.08	±0.04*
	4.7	*27.07.2011*	*1.3*		*0.9*		*0.13*		*0.13*	
Emitted mycotoxin load via			**14.5**		**18.8**		**0.62**		**0.78**	
Emitted fraction [%]			**0.24**		**0.45**		**0.10**		**0.12**	

nd: not detected; *triplicate measurements ± standard deviation of three measurements.

Figure S5.4. Occurrence of mycotoxins on the subplot cropped with spring wheat over the investigation period from June to October 2011 **A)** mycotoxin amounts in the cultivated plants, **B)** mycotoxin concentrations in the drainage water and the drainage water discharge.

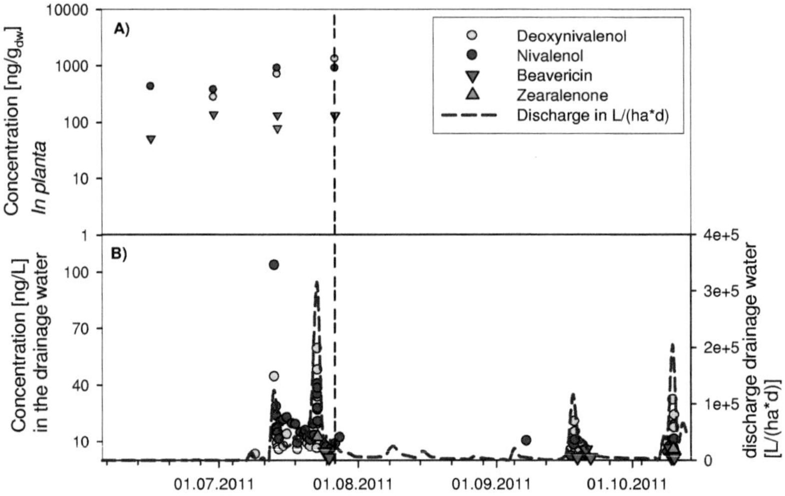

The dashed black line indicates the date of harvest (27.07.2011).

SI-5.7 Compilation of the median mycotoxin content in whole wheat plants at the time of harvest.

Table S5.8. Mean mycotoxin loads [g/ha] associated with the wheat plants at the day of harvest in 2010, and 2011.

		Type A		Type B trichothecenes					
upper field		HT-2	T-2	3-Ac-DON	DON	FUS-X	NIV	BEA	ZON
	29.07.2010	-	-	5.9	292	-	46	2.3	24.2
	14.07.2011	1.3	0.5	9.0	159	0.8	20	9.3	27.6
lower field									
	27.07.2011	-	-	-	6.1	-	4.2	0.6	0.6

-: no mycotoxins present at the time of harvest in the wheat plants.

SI-5.8 First flush event for individual rain events in drainage water samples and corresponding mycotoxin concentrations.

Figure S5.5. Precipitation (|), drainage water discharge (-- --) and the resulting mycotoxin concentrations (various symbols) in the drainage water during a precipitation event between 27 July – 1 August 2010. The left y-axis refers to the mycotoxin concentration, and the right y-axis to the drainage water discharge, as well as the precipitation.

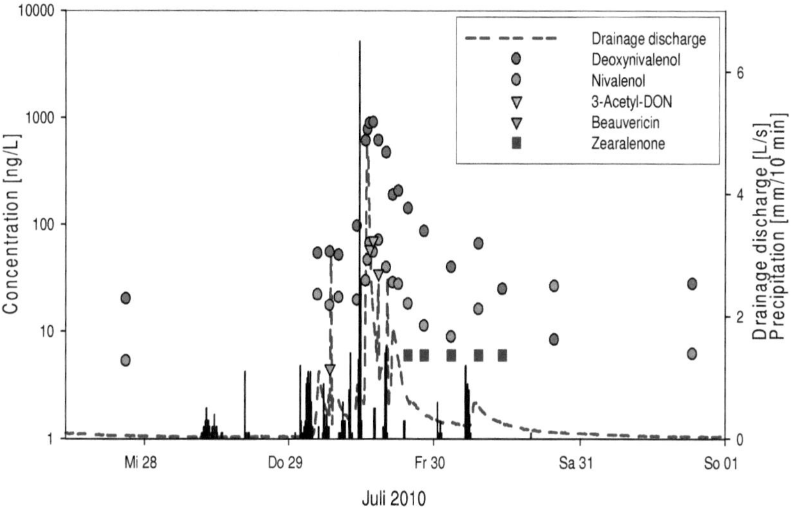

Reference (chapter 5 supporting information):

1. Hartmann, N., et al., Occurrence of zearalenone on *Fusarium graminearum* infected wheat and maize fields in crop organs, soil and drainage water. *Environ. Sci. Technol.* **2008,** *42*, 5455-5460.
2. FAO, *World reference base for soil resources.* 84 ed.; International Soil Reference and Information Centre: Viale delle Terme di Caracalla, 100 Rome, Italy, 1998.
3. Malek, E. and Bingham, G.E., Comparison of the bowen-ratio energy-balance and the water-balance methods for the measurement of evapotranspiration. *J. Hydrol.* **1993**, *146*, 209-220.

CHAPTER 6:

MYCOTOXINS IN THE ENVIRONMENT: II. OCCURRENCE AND ORIGIN IN SWISS RIVER WATERS

JUDITH SCHENZEL [1,2], KONRAD HUNGERBÜHLER[2], THOMAS D. BUCHELI[1]

[1]AGROSCOPE RECKENHOLZ-TANIKON, RESEARCH STATION ART, RECKENHOLZSTRASSE 191, CH-8046 ZURICH, SWITZERLAND
[2]INSTITUTE OF CHEMICAL AND BIOENGINEERING, SWISS FEDERAL INSTITUTE OF TECHNOLOGY, WOLFGANG-PAULI STRASSE 10, CH-8093 ZURICH, SWITZERLAND

REPRODUCED WITH PERMISSION FROM:

ENVIRONMENTAL SCIENCE AND TECHNOLOGY, 2012

VOLUME 46, ISSUE 24, PAGES 13076-13084

Abstract

Thirty-three different mycotoxins were surveyed over nearly two years in a typical Swiss waste water treatment plant (WWTP), as well as in Swiss midland rivers. Out of these, 3-acetyl-deoxynivalenol, deoxynivalenol (DON), nivalenol (NIV), and beauvericin (BEA), were detected. DON was quantified in all WWTP effluent grab samples with a maximum concentration of 73.4 ng/L, while the lowest concentration was observed for BEA with 1.3 ng/L. NIV was detected in about 37%, the other three compounds in 9–36% of the weekly or fortnightly integrated flow proportional river water samples. Concentrations were river discharge dependent, with higher numbers in smaller rivers, but mostly in the very low ng/L-range, with a maximum of 24.1, and 19.0 ng/L for NIV and DON, respectively. While NIV and DON prevailed in summer and autumn, BEA occurred mostly during winter. Summer and autumn seasonal load fractions were, however, not correlating with other river basin parameters indicative of the probably most obvious seasonal input source, i.e. *Fusarium graminearum* infected wheat crop areas. Nevertheless, together with WWTP effluents, these two sources largely explained the loads of mycotoxins quantified in river waters. The ecotoxicological relevance of mycotoxins as newly identified aquatic micropollutants has yet to be assessed.

6.1 Introduction

Mycotoxins are toxic metabolites produced by fungi of the genera *Alternaria*, *Aspergillus*, *Claviceps*, *Fusarium*, and *Penicillium*, which infect agricultural commodities and food worldwide. The most notorious compound classes are aflatoxins, ochratoxins, trichothecenes, resorcyclic acid lactones (RALs), fumonisins, and ergot alkaloids. They contaminate food, beverages and feed, and have been ranked as the most important chronic dietary risk factor, higher than synthetic contaminants, plant toxins, food additives or pesticide residues.[1] Expenses of up to billions of dollars per year worldwide are caused by mycotoxin mediated crop losses and human and animal health costs.[2,3] Thus, much research has been carried out on the analysis of mycotoxins in food and feed matrices, human and animal exposure assessment, and mycotoxin health effects.[2,4,5]

So far, the environmental exposure to mycotoxins has scarcely been studied. Recently published data on two prominent mycotoxins, i.e. deoxynivalenol (DON) and zearalenone (ZON), support the presumption that the aquatic environment can also be exposed to mycotoxins. Thus far, the identified main input sources of mycotoxins into the aquatic environment include 1) run-off and drainage water from fields cultivated with fungi-infected cereals, 2) husbandry animals (i.e., excretion from grazing livestock, run-off from feeding operations, or manure application), and 3) human excretion via sewer systems.

Mycotoxin emission from fields cultivated with wheat or corn by drainage water has recently been investigated in more detail. Hartmann et al.[6,7] reported occasional peak concentration up to 35 ng/L for ZON. Bucheli et al.[8] detected DON in concentrations between 23 ng/L and 4.9 µg/L, and explained the larger numbers compared to ZON with higher amounts in the contaminated crop, and its higher aqueous solubility. Schenzel et al.[9] observed three more mycotoxins, namely 3-acetyl-deoxynivalenol (3-AcDON), nivalenol (NIV), and beauvericin (BEA), by using a multi-residue screening method for various classes of mycotoxins.[10]

A second source for mycotoxins entering the aquatic environment is husbandry animals. Generally, mycotoxins are excreted by husbandry animals rather rapidly after the consumption of contaminated feed-stuff.[11] Up to 37% of the administered DON was excreted via urine in rats.[12] Up to 3 or 68% of DON or NIV, respectively, were detected in pig faeces,[13,14] whereas varying fractions of *α*- and *β*-zearalenol (ZOL), and ZON were recovered in cow faeces.[15] Besides the excretion of mycotoxins via grazing husbandry animals, mycotoxins enter the aquatic environment via run-off from animal feeding operations. Bartelt-Hunt et al.[16] reported concentrations of *α*-ZOL up to 5.2 µg/L in run-off from cattle feedlots. In this particular case, the detected ZON metabolites were emitted as a consequence of the application of the growth promoter *α*-zearalanol. Moreover, mycotoxins can still be present in manure. Hartmann et al.[17] found average ZON concentrations of 50 to 150 ng/g_{dw} in different types of manure. Schollenberger et al.[18] found DON in one of seven digested manure samples at the detection limit (20 µg/L), and deepoxy-DON was quantified between 6 and 20 µg/L in three of them. Hence, a mycotoxin input in rivers from husbandry animal waste seems likely. However, on an absolute scale, the emitted amounts are considerably lower than those from *Fusarium*-infected crops. For instance, the 50 to 150 mg/ha/y of ZON applied to agricultural areas in the form of manure[17] are significantly smaller than the roughly 30 g/ha/y produced in whole wheat plants.[6,9]

A third significant source of mycotoxins entering the aquatic environment is human excretion via sewer systems. The presence of mycotoxins in waste water treatment plants (WWTPs) in- and effluents was documented by several authors (reviewed in ref[19]). Wettstein et al.,[20] for example, reported influent concentrations for DON of 32 to 118 ng/L, and Lagana et al.[21] detected ZON and its metabolites *α*- and *β*-ZOL in concentrations from not detected to 18 ng/L. The compounds were mostly still present in respective effluents (DON: 19–79 ng/L[20]; ZON and *α*-/*β*-ZOL: not detected to 10 ng/L[21]), indicating their only partial removal by WWTPs. Trace levels of NIV, 3-AcDON, and BEA were recently also detected in WWTP effluents.[10]

The above-mentioned sources are probable causes for the exposure of surface waters with mycotoxins. Their occurrence, mostly in the low ng/L-concentration range, has been reported in various surface waters in Europe and the US.[8,21-23] In Switzerland, a previous study demonstrated the occurrence of DON and ZON with concentrations of up to 22 ng/L, and below the method quantification limit, respectively, in July and August 2007,[8] but no other seasons or mycotoxins were included in this study. To our knowledge, no information is available on the occurrence of any other mycotoxin in river water, except for some preliminary data on NIV, 3-AcDON and BEA.[10]

The aim of this study was to elucidate the occurrence and origin of a larger set of important mycotoxins in a representative WWTP effluent, as well as in Swiss river waters over nearly two years. Weekly and fortnightly integrated, flow-proportional concentrations in rivers from the Canton of Zurich and the Canton of Aargau will be presented. Influencing factors such as river discharge, seasonality, and water residence time will be discussed. The data will be correlated with parameters indicative of the presumed major input sources, such as agricultural areas cropped with small grain cereals (presented in detail in a companion manuscript[9]), and population equivalents in the catchments. Finally, the ecotoxicological relevance of the presence of these compounds in surface waters will be briefly addressed.

6.2 Materials and Methods

6.2.1 WWTP samples

Effluent grab samples (12L, n=1 for each sampling time point) were taken from the secondary sedimentation basin of the waste water treatment plant (WWTP) located at Kloten/Opfikon in the river Glatt valley in November 2009, May, and November 2010, April, July, and September 2011. For exact dates please refer to **Table 6.1**. The time points were selected to cover the different seasons of a year to obtain possible variations in the human food basket. Overall, around 50,000 inhabitants are connected to the WWTP Kloten/Opfikon. The

WWTP was chosen because it is representative for other WWTPs in Switzerland,[24] well investigated,[20,25,26] and only 2.5 kilometres downstream of the Office for Waste, Water, Energy, and Air (AWEL) river water sampling station Glatt at Oberglatt (D) (see **Figure 6.1**, **Table 6.3**).

6.2.2 River water samples

Several observation sites from the river monitoring programs of the AWEL and the Swiss government (National River Monitoring and Survey Program, NADUF) (Figure 1) were adapted to collect additional weekly (AWEL) and weekly/fortnightly (NADUF) integrated and flow proportional samples (1L, n=1 for each station and time point; for details see[27,28]). Samples were obtained from January 2010 to November 2011 from the following rivers and locations: Aa at Niederuster (A), Aabach at Mönchaltdorf B), Glatt (at Fällanden (C), Oberglatt (D), and Rheinsfelden (E)), and Töss (at Rämismühle (F), and Freienstein (I)), Kempt at Winterthur (G), Eulach at Wülflingen (H) (all located in the Canton of Zurich, and belonging to the AWEL monitoring program), Thur (J) at Andelfingen (Canton of Zurich), Aare (L) at Brugg, Rhine (K) at Rekingen (both located in the Canton of Aargau) (all part of NADUF). For details see **Figure 6.1**.

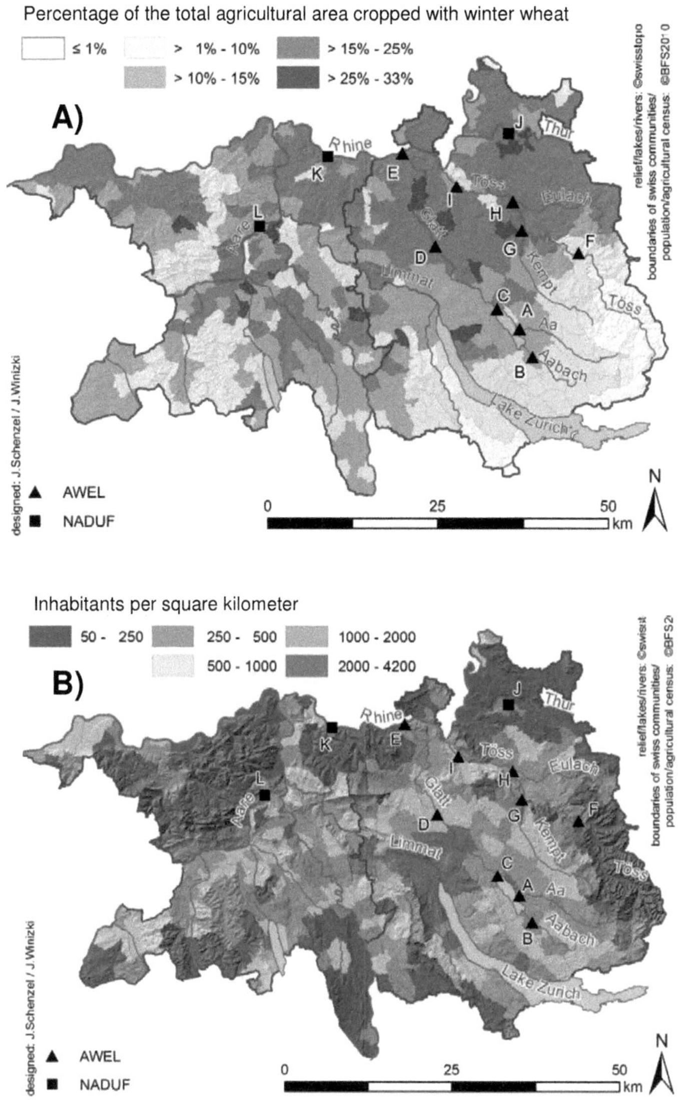

Figure 6.1: Map of the Canton of Zurich and the Canton of Aargau, including the location of the river water sampling stations in the river Glatt (A-E), Töss (F-I); all from the AWEL monitoring program (▲), Thur (J), Rhine (K), and Aare (L); all from the NADUF monitoring program (■). Further details about the sampling stations are provided in Table 2. The colored areas in panel A and B indicate the fraction of agricultural area cropped with winter wheat [%] per county, and the inhabitants per square kilometre (Source: Swiss Federal Statistical Office; Statistic of the agricultural area of Switzerland for the year 2010), respectively.

6.2.3 Analyses and data interpretation

All mycotoxins were quantified as described in Schenzel et al.[10] Briefly, the target compounds were solid-phase extracted from filtered water samples (1 L) with Oasis HLB cartridges (6 mL, 200 mg; Waters Corporation, Milford MA, USA). Analytes were eluted with 5 mL of methanol, further reduced under a nitrogen gas stream to 100±5 µL, reconstituted in 900 µL of Milli-Q water/methanol (90:10, v/v), and transferred into 1.5-mL amber glass vials. The samples were stored at +4°C and analyzed within 48 h. LC-MS/MS was performed on a Varian 1200L LC-MS instrument (VarianInc, Walnut Creek CA, USA). The mycotoxins were separated using a 125 mm x 2.0 mm i.d., 3 µm, Pyramid C_{18} column, with a 20 mm x 2.1 mm i.d., 3 µm guard column of the same material (Macherey-Nagel GmbH & Co. KG, Düren, Germany) and two chromatographic runs were used for the detection of all compounds, one followed by positive and one by negative ionization mode. Details about the analytical method and its quality control are also given in the **SI**. Positive identification criteria for the target analytes were their relative retention time, and their main-to-second product ion ratio. If an analyte was detectable but not quantifiable, its concentration was set equal to its limit of detection (LOD). Mycotoxin loads in the WWTP effluent were calculated from the quantified concentrations and the effluent water discharges, which were obtained from the WWTP protocols (personal communication C. Liebi, executive director WWTP Kloten/Opfikon). Mycotoxin loads in river water samples were calculated from the quantified concentrations and the mean weekly or fortnightly river water discharge, which were obtained from the sampling station protocols.[29,30] Seasonal load fractions (SLF) for each sampling station and detected mycotoxin were calculated from total mycotoxin load quantified in summers and autumns, divided by the one over the whole investigation period, in percent. Quantification of uncertainties is described where they appear in text and table footnotes.

6.3 Results and Discussion

Based on the mycotoxin sources outlined in the introduction, in principle, each of the 33 compounds included in the analytical method could potentially be present in WWTP or river water samples. Interestingly, only four out of the 33 mycotoxins, namely 3-AcDON, DON, NIV, and BEA (all classified as field produced mycotoxins[31]), were detected in WWTP effluents. Reasons for the absence of all others in WWTP effluents may be their overall lower absolute human intake, particularly in the case of storage-mycotoxins such as aflatoxin B_1 or ochratoxin A[32,33] (reviewed in ref[19]), their metabolization,[34] or sorption to sludge and consequently removal from the aqueous phase due to higher Log K_{ow} values[35] (**Table 6.2**). The four mycotoxins emitted from WWTP belonged to those that were also found in drainage water samples from agricultural runoff (see companion paper[9]). They were also the only ones detected in the river water samples investigated here. Therefore, from here on we will restrict the discussion to these regularly observed compounds.

Table 6.1: Occurrence of mycotoxins in the effluent of the secondary sedimentation basin of the WWTP Kloten/Opfikon.

	Water	Type B trichothecenes							
Date	**discharge**	**3-AcDON**		**DON**		**NIV**		**BEA**	
	[m3/d]	[ng/L]	[µg/cap/d]	[ng/L]	[µg/cap/d]	[ng/L]	[µg/cap/d]	[ng/L]	[µg/cap/d]
23.11.2009	12460	14.6	3.6	32.9	8.1	nd.	nd.	1.3	0.3
19.05.2010	16790	20.6	6.9	19.2	6.4	nd.	nd.	1.3	0.3
09.11.2010	12950	42.0	1.5	73.4	18.9	1.6	0.4	1.3	0.3
13.04.2011	13340	6.0	1.6	48.5	12.8	9.6	2.5	1.3	0.3
19.07.2011	20910	23.0	9.6	38.7	16.1	38.8	16.1	nd.	nd.
17.10.2011	18400	19.4	7.1	33.9	12.4	9.8	4.1	nd.	nd.
min		6.0	1.5	19.2	6.4	1.6	0.4	1.3	0.3
max		42.0	9.6	73.4	18.9	38.8	16.1	1.3	0.3
median		20.0	5.2	36.3	12.6	9.7	3.3	1.3	0.3

nd.: not detected

Table 6.2: Physical-chemical properties of the mycotoxins regularly quantified in river water.

	Type B trichothecenes			
	3-AcDON	DON	NIV	BEA
Molecular structure	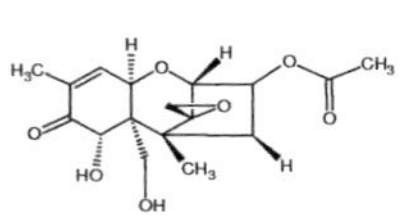	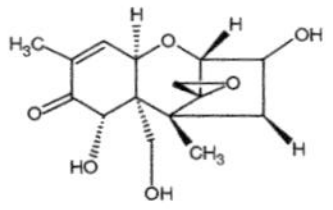	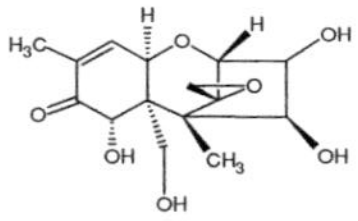	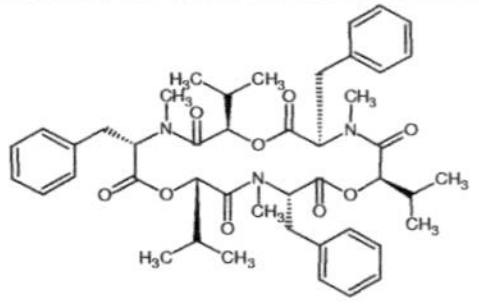
CAS number	50722-38-8	51481-10-8	23282-20-4	26048-05-5
Molecular formula	$C_{17} H_{22} O_7$	$C_{15} H_{20} O_6$	$C_{15} H_{20} O_7$	$C_{45} H_{57} N_3O_9$
Molecular weight [g/mol]	338.4	296.3	312.3	783.9
Log K_{ow}	$1.3^{a,b}$	$0.1^{a,b}$	$0.0^{a,b}$	5.5^{c}
Log K_{oc}	$< 0.7^{b}$	$< 0.7^{b}$	$< 0.7^{b}$	$5.9^{a,b}$
pK_a [d]	11.8	11.9	11.8	-
Half-life in natural waters [d][c]	55	38	38	180
Produced by *Fusarium* species[e]	*F. graminearum*	*F. graminearum*	*F. poae*, *F. graminearum*, *F. crookwellense*	*F. avenaceum*, *F. poae*

[a] Calculation with COSMOtherm. [b] From ref.[35] [c] Calculation with EPISuite v4.0. [d] Calculation with SPARC v4.5. [e]From ref.[36]
-, no data available.

6.3.1 Mycotoxin concentrations and loads in WWTP effluents

As mentioned above, four compounds, i.e., three type B trichothecenes (3-AcDON, DON, and NIV), and BEA, were detected in WWTP grab samples from the secondary sedimentation basin in the facility Kloten/Opfikon (**Table 6.1**). The type B trichothecenes were by far the most prominent representatives in terms of concentrations. DON showed the highest concentration of 73 ng/L, and was followed by 3-AcDON and NIV with 42, and 39 ng/L, respectively. Concentrations observed for BEA were generally low and ranged from not detected to 3.4 ng/L. The here reported DON concentrations in the WWTP effluent are in good accordance with data presented by Wettstein & Bucheli.[20] For the other three compounds (NIV, 3-AcDON, BEA) only our own initial data is available so far,[10] but it can be presumed from the structural and chemical-physical similarity of the type B trichothecenes, e.g. in terms of hydrophobicity[35] and functional groups (**Table 6.2**), that 3-AcDON and NIV exhibit equally low elimination rates in WWTPs as observed for DON.[20] This corroborates the importance of WWTP as a source of this compound class in surface waters. While ZON and its metabolites α- and β-ZOL were present in Italian WWTP effluents in concentrations <10 ng/L,[21] such concentrations were below the method detection limit of the here applied multi-residue screening method.[10]

Together with the only slightly variable water discharge (**Table 6.1**), and considering the 50,000 inhabitants attached to this WWTP, the quantified concentrations translate into daily loads between 320 mg/d and 950 mg/d, or 6.4 to 18.9 μg per capita per day, respectively, in the case of DON (**Table 6.1**). These numbers match excellently with those reported by Wettstein & Bucheli (340–1010 mg/d, and 6.7–20.1 μg/capita/d).[20] Analogously, average loads of 5.8±7.0, 6.6±3.5, and 0.3 μg/cap/d were obtained for NIV, 3-AcDON, and BEA, respectively (**Table 6.1**).

6.3.2 Mycotoxin concentrations in AWEL and NADUF rivers

DON, NIV, 3-AcDON, and BEA were regularly detected in the rivers Glatt (incl. tributaries), Töss (incl. tributaries), Thur, and still occasionally in the considerably larger rivers Aare and Rhine throughout the investigation period from January 2010 to October 2011 (**Table 6.3**).

DON and NIV were by far the most prominent mycotoxins in terms of frequency of occurrence (390 and 403 detects out of 1054 analyses, i.e. 36 and 37%, respectively, **Tables 6.3** and **S6.4**). 3-AcDON and BEA were detected in 12 and 9% of all analyses (**Tables S6.5** and **S6.6**). The median concentrations of DON in all investigated rivers were comparable and ranged from 3.1–7.1 ng/L (**Table 6.3**). Apart from NIV, which exhibited comparable median concentrations of detected to 7.6 ng/L, the other two compounds were mostly below LOQ (**Tables S6.4-S6.6**). The maximum concentrations detected for DON, NIV, 3-AcDON and BEA were 19.0, 24.1, 19.2, and 22.0 ng/L, respectively (**Tables 6.3** and **S6.4-S6.6**).

The mean river discharge of the investigated rivers covered a wide range from 0.97–9.7 m^3/s for AWEL, and from 53–468 m^3/s for NADUF rivers, respectively, within the nearly two years of investigation (**Table 6.3**). Generally, rivers with higher water discharge tended to exhibit lower mycotoxin concentrations than rivers with smaller discharges (**Figure S6.4**). For DON maximum concentrations of 16.3-19 ng/L were found in smaller rivers with median water discharges between 0.97 and 1.6 m^3/s (**Table 6.3**). Additionally, concentrations were lower at times of heavy rain events. Both observations point to dilution as one of the main determinant of mycotoxin concentrations. While the occurrence of DON, 3-AcDON, and BEA was similar in 2010 and 2011 in all rivers, NIV concentrations were generally higher in 2011 than 2010 (**Figures 6.2** and **S6.6-S6.12**). Apparently, NIV producing *Fusarium* species such as *F. poae* and *F. crookwellense* prevailed in the corresponding river catchments in 2011.

Table 6.3: Concentrations of Deoxynivalenol [ng/L] in the weekly and fortnightly collected flow-proportional surface water samples from the AWEL and NADUF sampling sites from January 2010 to November 2011.

River location	#detected/ #samples analyzed	min concentration [ng/L][c]	max concentration [ng/L][c]	median concentration [ng/L][c]	mean river discharge $[m^3/s]$[d]	cumulative DON load [kg][d]	wheat in river catchment area $[km^2]$[e]	% of WWTP effluent at sampling site[f]	Inhabitants in river catchment	SLF[g] [%]
River Glatt[a]										
Aa at Niederuster (A)	41/95	det.	16.6	7.1	1.88	0.4	3.3	15	$7.0*10^4$	75
Aabach at Mönchaltorf (B)	51/95	det.	19	6.5	1.14	0.3	2.1	10	$1.5*10^4$	90
Glatt at Fällanden (C)	21/95	det.	14	5.8	4.66	0.3	6.6	15	$1.2*10^5$	92
Glatt at Oberglatt (D)	51/95	det.	12.6	7.1	6.44	1.1	13.2	19	$2.1*10^5$	64
Glatt at Rheinsfelden (E)	57/95	det.	14.7	5.9	8.43	1.5	20.3	20	$2.8*10^5$	78
River Töss[a]										
Töss at Rämismühle (F)	14/95	det.	10.8	4.2	3.76	0.1	1	0.7	$9.4*10^3$	70
Kempt at Winterthur (G)	37/95	det.	17.2	5.3	1.58	0.2	4	13	$2.4*10^4$	84
Eulach at Wülflingen (H)	45/95	det.	16.3	5.2	0.97	0.1	13	4	$6.3*10^3$	70
Töss at Freienstein (I)	42/95	det.	11.5	5.3	9.7	1.1	30.8	10	$1.8*10^5$	68
Larger Rivers in the Swiss Midlands[b]										
Thur at Andelfinden (J)	23/78	det.	11	5.1	52.8	4.9	11.3	0.1	-	80
Rhine at Rekingen (K)	Apr 73	det.	11.5	4.7	468	10.9	88.1	-	-	100
Aare at Brugg (L)	Apr 48	det.	5.9	3.1	285	3.1	34.4	-	-	100

[a] monitoring network: AWEL (Office for waste, water, energy and air, Canton Zurich, Switzerland); [b] monitoring network: NADUF (National River Monitoring and Survey Program); Letter in parenthesis indicate abbreviations in **Figure 6.1**. [c] out of detected samples. [d] after 22 month. Det: detected (above LOD (1.5. ng/L), but below LOQ (4.5 ng/L)). [e] Source: Statistic of the agricultural area of Switzerland for the year 2010. [f] Source: AWEL (Office for waste, water, energy and air, Canton Zurich, Switzerland); [g] SLF: seasonal load fraction, sum of the DON loads detected in summer and autumn to the total cumulative DON load at each specific station. - : no information available.

6.3.3 Mycotoxin loads in AWEL and NADUF rivers

Cumulative DON loads of 0.38±0.03 and 0.25±0.03 kg were obtained during the 22 months of investigation in Aa (A) and Aabach (B), respectively, (**Table 6.3**), the two tributaries of lake Greifen (**Figure 6.1**A & B). At the outflow of the lake Greifen (Glatt at Fällanden (C)) the cumulative load of DON was smaller than the load which entered the lake (**Table 6.3**). This was somewhat expected, considering the mean epilimnion water residence time of 140 days,[36] and an estimated DON half-life of 38 d in natural waters (EPISuite v4.0, and own unpublished data). Further downstream at the river Glatt, the cumulative loads of DON were additive. The same additive behavior was observed in the river Töss and its tributaries (**Table 6.3**). The highest cumulative DON load of 10.9±1.0 kg was recorded at the Rhine in Rekingen (K). Generally, a similar additive behavior of cumulative loads was observed for NIV, 3-AcDON, and BEA in river Glatt and Töss (**Tables S6.4-S6.6**). Losses in lake Greifen were less pronounced though for 3-AcDON and BEA (**Figures S6.5-S6.7**), which is in accordance with their higher half-lifes (**Table 6.2**).

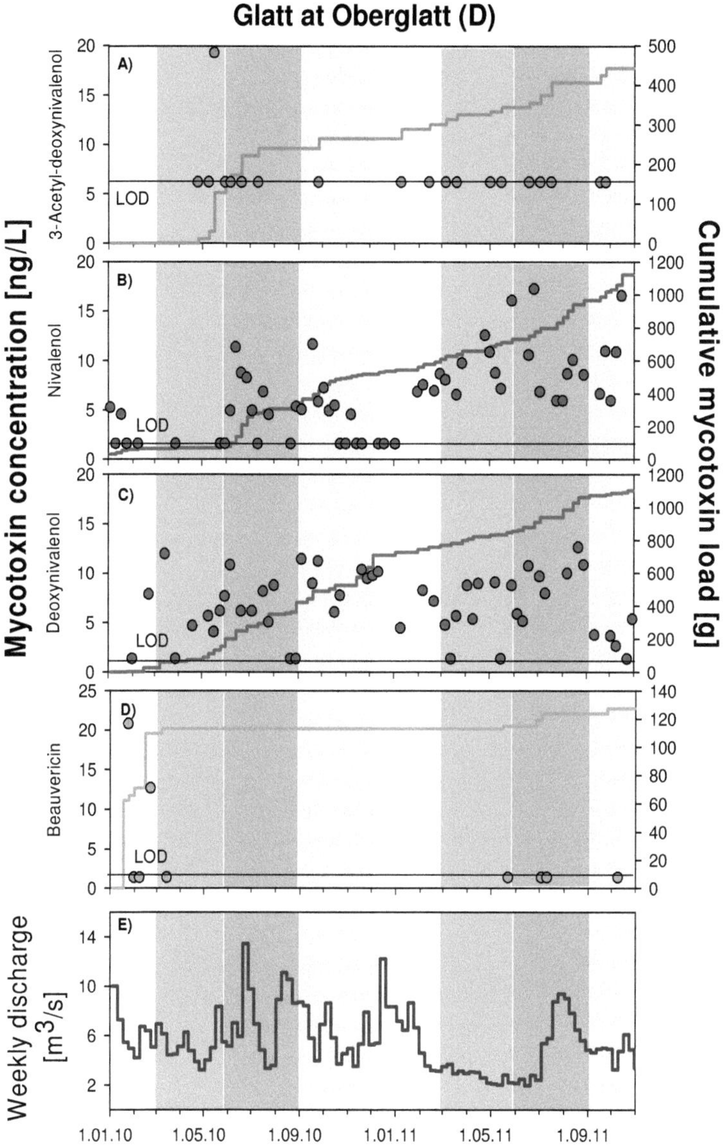

Figure 6.2: Occurrence of mycotoxins in river Glatt at Oberglatt (station D) from January 2010 until October 2011: 3-acetyl-deoxynivalenol (A); nivalenol (B); deoxynivalenol (C); beauvericin (D); weekly river water discharge [m^3/s] (E). The left y-axes in panel A-D indicate concentrations in [ng/L], and the weekly river water discharge [m^3/s] (E), respectively. The right y-axes in panel A-D indicate the cumulative load [g]. The colored stripes indicate the seasons: spring: green; summer: red; autumn: yellow; winter: white. LOD: limit of detection; points under LOD-level: not detected.

6.3.4 Seasonal load fractions of mycotoxins

The temporal occurrences of the four detected mycotoxins in the river Glatt at Oberglatt (D) are depicted in **Figure 6.2**A-D, and for all other monitoring stations in **Figures S6.5-S6.12**. In light of the fact that mycotoxins are emitted from *Fusarium*-infected small grain cereals seasonally,[8,37] and from WWTP continuously, we determined their relative importance by calculating SLF for each of the four mycotoxins and for each sampling station (**Table 6.3**, and **S6.4-S6.6**). Correspondingly, SLF of DON, and NIV in ref[9] was 97 and 89%, respectively, whereas mycotoxins are taken in and excreted by humans more or less constantly over the whole year (own data presented above, and ref[20]). Apparently, with SLFs of DON and NIV between 60±4 and 100±1%, the occurrence of these mycotoxins exhibited a strong seasonal dynamic (**Tables 6.3** and **S6.4**). Interestingly, not all type B trichothecenes showed the highest seasonality at the same station. For DON in particular, a clear seasonality (84±3%≤SLF≤92±2%) was observable at Aabach at Mönchaltorf (B), Glatt at Fällanden (C), and Kempt at Winterthur (G) (**Table 6.3**, **Figures S6.9**, **S6.10**, and **S6.13**). In contrast, seasonality for NIV was clearly observable at Eulach at Wülfingen (H) and Töss at Rämismühle (F) (SLF=83±5%, and 80±2%, respectively). These differences in seasonality are probably caused by the spatial variability of different *Fusarium* species, and their infection rates. Overall, the situation was less clear for 3-AcDON, and BEA, because of a high frequency of no detects. Nevertheless, with SLFs in AWEL stations between 8±1 and 51±2% (**Table S6.6**), BEA exhibited a clearly different seasonality. Note that a similar difference in seasonal occurrence between trichothecenes and BEA was also observed in our companion paper on emission[9] and attributed to the higher hydrophilicity and mobility of the formers.

Generally, the relative importance of *Fusarium*-infected crops and WWTP as sources of mycotoxins are expected to be higher and lower, respectively, in rural, and *vice versa* in urban regions. Correspondingly, the ratio of numbers representative for each of these sources, i.e. the wheat cultivation area, and the number of inhabitants, in a given catchment (**Tables 6.3**) should be indicative for the relative importance of *Fusarium*-infected crops, and positively correlated with the SLF. This was, however, not the case for any of the four mycotoxins (see **Figure S6.13**A-D). Thus, we inspected the relationship of cumulative mycotoxin loads from the nine locations of the rivers Glatt and Töss with number of inhabitants, and the wheat cultivation area, respectively, in more detail by a regression analysis (**Table S6.7**). Regressions were consistently better (higher R^2, lower p-values) with the former, both when considering the whole period of investigation and summer-autumn seasons only. Even when the total agricultural area in each catchment was considered, the regressions were still consistently better (higher R^2, lower p-values) for the number of inhabitants (data not shown). Additionally, the non-linearity of this relationship was also tested, but without any success.

Conclusively, although mycotoxin loads in river waters mostly showed a clear seasonal trend, this trend could not be attributed to a quantitative parameter for its most obvious source. Out of the many possible reasons, one may be that we could not account for the third possible (and likely also seasonal) source of mycotoxins, i.e., emission from husbandry animal manure applications during the spring to autumn, for which no reliable numbers are available this far (see above).

6.3.5 Plausibility test; estimated contributions from the presumed main sources

As outlined above, two main input sources of mycotoxins into the aquatic environment are, on the one hand, human excretions and WWTP in- and effluents,[21,22] and, on the other hand, run-off from agricultural areas cropped with small grain cereals, like wheat or corn.[6,9,10,20,23] In the following, we attempt to predict mycotoxins loads emitted from these sources and compare them with measured ones in river water.

Using the average ± standard deviation per capita emissions (**Table 6.1**) and the total number of inhabitants connected to each sampling station via

WWTP (**Table 6.3**), we predicted amounts of mycotoxins originating from human emission via WWTP (**Figures 6.3** and **S6.14-S6.16**, **Table S6.8**). The estimated loads for the whole investigation period from WWTPs ranged from 0.05±0.02–1.5±0.6 kg, 0.01±0.01–0.4±0.5 kg, 0.02±0.01–0.6±0.3 kg, and 4±1–108±15 g for DON, NIV, 3-AcDON, and BEA, respectively, with the minimum load in the Eulach at Wülfingen (H) and the maximum load in the Töss at Freienstein (I) (**Table S6.8**).

Similarly, we can roughly estimate weekly mycotoxin river water loads due to the emission from *Fusarium graminearum.* infected crop fields, using data from the companion paper[9], and agricultural areas cropped with winter wheat in the catchment of each of the river water sampling station (**Table 6.3**). An experimental field with an averaged *F. graminearum* infection rate of 94±8% (infection data from ref[9] determined according to seed health test[38]) contained on average over two seasons as much as 228±93, 35±15, and 8±2 of DON, NIV, and 3-AcDON, respectively, per hectare per year, out of which some 0.075±0.064, 0.08±0.01, 0.06±0.07% were emitted via drainage water from December 2009 to November 2011.[9] In 2007, and 2008, the average infestation of wheat with *F. graminearum* in Switzerland was 4.9 and 8.7%, respectively.[39] Using an averaged infection percentage of 6.8±2.7%, and assuming similar mycotoxin production and emission rates for the wheat areas in the catchments as from our experimental field, we can calculate estimated mycotoxin loads for each sampled river water station of the AWEL monitoring program. *F. graminearum* is not a producer of beauvericin, accordingly no load estimation was possible. Over the whole investigation period, DON, NIV, and 3-AcDON are estimated to range between 0.6±0.9–1740±3020, 0.4±0.6–720±1110, and 0.05±0.08–231±368 g, respectively, with the minimum load in the Glatt at Fällanden (C) and the maximum load in the Glatt at Rheinsfelden (E) (**Figures 6.3** and **S6.14-S6.16**, **Table S6.8**).

Equipped with these simple proxy tools, we can compare the estimated mycotoxins amounts at any river sampling site with those actually measured. For example, at the Glatt at Oberglatt (D), four WWTP discharge their treated waste water from $1.0*10^5$ inhabitants into the river, making up in total 19% of the average water flow at this site,[40] and 13.2 km^2 of the area in the river catchment is cropped with wheat (**Table 6.3**). Hence, we estimate human emissions over the period of investigation to sum up to 0.8±0.3 kg DON, and estimated emissions from wheat crops to contribute another 0.5±1.0 kg (**Figure 6.3**). The total of 1.4±1.3 kg of DON corresponds well to the detected total DON load at this location of 1.1±0.1 kg (Figure 3). Hence, the emission from wheat crops accounts just for a half of the total DON load in the Glatt at Oberglatt. Further downstream, at the next AWEL station Glatt at Rheinsfelden (E), the number of WWTP (8) and inhabitants ($1.6*10^5$) in the river catchment has increased, as well as the total area cropped with wheat of 20.3 km^2, and the total quantified cumulative DON load was 1.5±0.1 kg (see **Table 6.3**). Here, estimated 1.4±0.5 kg and 1.7±3.0 kg DON were emitted from WWTP effluents and agricultural areas cropped with wheat, respectively, over the whole investigation period (**Figure 6.3**). Overall, although the considerable uncertainties of our predictions were limiting the actual source apportionment, the fact that averaged estimated and measured numbers matched within a factor of about two is rather satisfactory. Similar matches between predicted and observed mycotoxin loads were achieved for the NIV and 3-AcDON, whereas the numbers deviated somewhat more for BEA (**Figures S6.14-S6.16**).

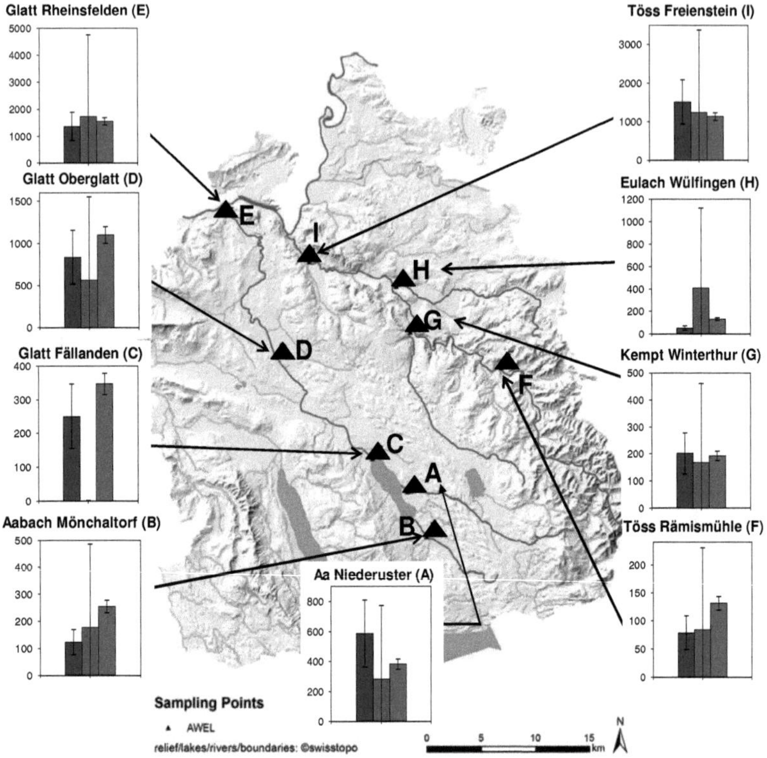

Figure 6.3: Surface waters of the Canton of Zurich with the river Glatt and Töss highlighted in dark blue. Inserted bar charts show the estimated amounts from WWTP effluents [g] (blue), and from agricultural areas cropped with winter wheat [g] (green). They are compared to the total deoxynivalenol load [g] (red) at each sampling station over an investigation period from January 2010 until October 2011.

6.3.6 Ecotoxicological relevance

Compared to mycotoxin concentrations at source, i.e., in drainage water,[9] the concentrations observed in river water are up to two orders of magnitude lower, obviously because of dilution. Nevertheless, in terms of exposure frequency and concentrations, mycotoxins are not unlike other aquatic micropollutants and should be considered as such.[8] The ecotoxicological relevance of the frequent (DON, NIV) or occasional (BEA, 3-AcDON) presence of mycotoxins in surface waters in the low nanogram per liter range may be low, but is still unknown, mainly because effect studies (summarized in our companion paper[9]) with these mycotoxins are largely lacking.

Acknowledgements

We thank the Swiss Federal Office for the Environment for the financial support. We gratefully acknowledge P. Niederhauser (AWEL) and B. Luder (NADUF) for sample donation. Additionally, we would like to acknowledge Pakeerathan Srikanthan and Stefan Weissen for their excellent help with the SPE extractions of the aqueous samples. Further, we would also like to thank J. Winizki (ART) for the GIS-maps, and S. Vogelgsang, and H.-R. Forrer for their helpful comments on the manuscript.

References (chapter 6)

1. Kuiper-Goodman, T., Food safety: Mycotoxins and phycotoxins in perspective. In *Mycotoxins and Phycotoxins - Developments in Chemistry, Toxicology and Food Safety*, Miraglia, M.; VanEgmond, H. P.; Brera, C.; Gilbert, J., Eds. Alaken, Inc: Ft Collins, 1998; pp 25-48.
2. Hussein, H. S.; Brasel, J. M., Toxicity, metabolism, and impact of mycotoxins on humans and animals. *Toxicology* **2001,** *167*, 101-134.
3. Nganje, W. E.; Bangsund, D. A.; Leistritz, F. L.; Wilson, W. W.; Tiapo, N. M., Regional economic impacts of Fusarium Head Blight in wheat and barley. *Rev. Agric. Econ.* **2004,** *26*, 332-347.
4. Krska, R.; Welzig, E.; Boudra, H., Analysis of Fusarium toxins in feed. *Ani. Feed Sci. Technol.* **2007,** *137*, 241-264.
5. Shephard, G. S., Determination of mycotoxins in human foods. *Chem. Soc. Rev.* **2008,** *37*, 2468-2477.
6. Hartmann, N.; Erbs, M.; Wettstein, F.E.; Schwarzenbach, R.P.; Bucheli, T. D., Occurrence of zearalenone on *Fusarium graminearum* infected wheat and maize fields in crop organs, soil and drainage water. *Environ. Sci. Technol.* **2008,** *42*, 5455-5460.
7. Hartmann, N.; Erbs, M.; Wettstein, F. E.; Schwarzenbach, R. P.; Bucheli, T. D., Quantification of estrogenic mycotoxins at the ng/L level in aqueous environmental samples using deuterated internal standards. *J. Chromatogr. A* **2007,** *1138*, 132-140.
8. Bucheli, T. D.; Wettstein, F. E.; Hartmann, N.; Erbs, M.; Vogelgsang, S.; Forrer, H.-R.; Schwarzenbach, R. P., Fusarium Mycotoxins: Overlooked Aquatic Micropollutants? *Journal of Agriculture and Food Chemistry* **2008,** *56*, (3), 1029-1034.
9. Schenzel, J.; Hungerbuhler, K.; Bucheli, T. D., Mycotoxins in the environment: I. Production and emission from an agricultural test field. *Environ. Sci. Technol.* **submitted.**
10. Schenzel, J.; Schwarzenbach, R. P.; Bucheli, T. D., A multi residue screening method to quantify mycotoxins in aqueous environmental samples. *J. Agric. Food Chem.* **2010,** *58*, 11207-11217.

11. Cavret, S.; Lecoeur, S., Fusariotoxin transfer in animal. *Food Chem. Toxicol.* **2006,** *44*, 444-453.

12. Meky, F. A.; Turner, P. C.; Ashcroft, A. E.; Miller, J. D.; Qiao, Y. L.; Roth, M. J.; Wild, C. P., Development of a urinary biomarker of human exposure to deoxynivalenol. *Food Chem. Toxicol.* **2003,** *41*, 265-273.

13. Danicke, S.; Valenta, H.; Doll, S., On the toxicokinetics and the metabolism of deoxynivalenol (DON) in the pig. *Arch. Ani. Nutrit.* **2004,** *58*, 169-180.

14. Hedman, R.; Pettersson, H.; Lindberg, J. E., Absorption and metabolism of nivalenol in pigs. *Arch. Ani. Nutrit.* **1997,** *50*, 13-24.

15. Seeling, K.; Danicke, S.; Ueberschar, K. H.; Lebzien, P.; Flachowsky, G., On the effects of Fusarium toxin-contaminated wheat and the feed intake level on the metabolism and carry over of zearalenone in dairy cows. *Food Addit. Contam.* **2005,** *22*, 847-855.

16. Bartelt-Hunt, S. L.; Snow, D. D.; Kranz, W. L.; Mader, T. L.; Shapiro, C. A.; Donk, S. J. v.; Shelton, D. P.; Tarkalson, D. D.; Zhang, T. C., Effect of growth promotants on the occurrence of endogenous and synthetic steroid hormones on feedlot soils and in runoff from beef cattle feeding operations. *Environ. Sci. Technol.* **2012,** *46*, 1352-1360.

17. Hartmann, N.; Erbs, M.; Wettstein, F. E.; Hoerger, C. C.; Schwarzenbach, R. P.; Bucheli, T. D., Quantification of zearalenone in various solid agroenvironmental samples using D-6-zearalenone as the internal standard. *J. Agric. Food Chem.* **2008,** *56*, 2926-2932.

18. Schollenberger, M.; Muller, H. M.; Rufle, M.; Suchy, S.; Dejanovic, C.; Frauz, B.; Oechsner, H.; Drochner, W., Simultaneous determination of a spectrum of trichothecene toxins out of residuals of biogas production. *J. Chromatogr. A* **2008,** *1193*, 92-96.

19. Bester, K.; McArdell, C. S.; Wahlberg, C.; Bucheli, T. D., Quantitative mass flows of selected xenobiotics in urban waters and waste water treatment plants. In *Enviornmental Pollution* Fatta-Kassinos, D.; Bester, K.; Kuemmerer, K., Eds. Springer Netherlands: Dordrecht, 2010; Vol. 16.

20. Wettstein, F. E.; Bucheli, T. D., Poor elimination rates in waste water treatment plants lead to continuous emission of deoxynivalenol into the aquatic environment. *Wat. Res.* **2010,** *44*, 4137-4142.

21. Lagana, A.; Bacaloni, A.; De Leva, I.; Faberi, A.; Fago, G.; Marino, A., Analytical methodologies for determining the occurrence of endocrine disrupting chemicals in sewage treatment plants and natural waters. *Anal. Chim. Acta* **2004,** *501*, 79-88.
22. Gromadzka, K.; Waskiewicz, A.; Golinski, P.; Swietlik, J., Occurrence of estrogenic mycotoxin - Zearalenone in aqueous environmental samples with various NOM content. *Wat. Res.* **2009,** *43*, 1051-1059.
23. Kolpin, D. W.; Hoerger, C. C.; Meyer, M. T.; Wettstein, F. E.; Hubbard, L. E.; Bucheli, T. D., Phytoestrogens and mycotoxins in Iowa streams: An examination of underinvestigated compounds in agricultural basins. *Journal of Environmental Quality* **2010,** *39*, 2089-2099.
24. Huset, C. A.; Chiaia, A. C.; Barofsky, D. F.; Jonkers, N.; Kohler, H. P. E.; Ort, C.; Giger, W.; Field, J. A., Occurrence and mass flows of fluorochemicals in the Glatt Valley watershed, Switzerland. *Environ. Sci. Technol.* **2008,** *42*, 6369-6377.
25. Hoerger, C. C.; Schenzel, J.; Strobel, B. W.; Bucheli, T. D., Analysis of selected phytotoxins and mycotoxins in environmental samples. *Anal. Bioanal. Chem.* **2009,** *395*, 1261-1289.
26. McArdell, C. S.; Molnar, E.; Suter, M. J. F.; Giger, W., Occurrence and fate of macrolide antibiotics in wastewater treatment plants and in the Glatt Valley Watershed, Switzerland. *Environ. Sci. Technol.* **2003,** *37*, 5479-5486.
27. Davis, J. S.; Fahrni, H.-P.; Liechti, P.; Spreafico, M.; Stadler, K.; Zobrist, J., Das nationale Programm für die analytische Daueruntersuchung der schweizerischen Fliessgewässer -eine Standardbestimmung. *Gas Wasser Abwasser* **1985,** *65*, 123-135.
28. Känel, B.; Niederhauser, P.; Meier, W. Zustand der Fliessgewässer in den Einzugsgebieten von Sihl, Limmat und Zürichsee; Messkampagne 2006 / 2007; AWEL: Zurich, 2008; pp 1-62.
29. AWEL http://www.awel.zh.ch/internet/baudirektion/awel/de/wasserwirtschaft/gewaesserqualitaet/fg_qualitaet.html.
30. NADUF National River Monitoring and Survey Programme. http://www.bafu.admin.ch/hydrologie/01831/01840/10435/index.html?lang=de

31. Scudamore, K. A.; Livesey, C. T., Occurrence and significance of mycotoxins in forage crops and silage: a review. *J. Sci. Food Agric.* **1998,** *77*, 1-17.
32. JECFA, Safety evaluation of certain mycotoxins in food. Deoxynivalenol, HT-2 and T-2 toxin. http://www.inchem.org/documents/jecfa/jecmono/v47je01.htm.
33. SCOOP *Collection of occurrence data of Fusarium toxins in food and assessment of dietary intake by the population of the EU member states.* ; SCOOP TASK 3.2.10. Directorate-General Health and Consumer Protection: 2003.
34. Wu, Q. H.; Dohnal, V.; Huang, L. L.; Kuca, K.; Yuan, Z. H., Metabolic pathways of trichothecenes. *Drug Metab. Rev.* **2010,** *42*, (2), 250-267.
35. Schenzel, J.; Goss, K.-U.; Schwarzenbach, R. P.; Bucheli, T. D.; Droge, S. T. J., Experimentally determined soil organic matter-water sorption coefficients for different classes of natural toxins and comparison with estimated numbers. *Environ. Sci. Technol.* **2012,** *46*, 6118-6126.
36. Schwarzenbach, R. P.; Gschwend, P. M.; Imboden, D. M., *Environmental organic chemistry.* John Wiley & Sons, Inc.: New York, 2003; Vol. second edition.
37. Hartmann, N.; Erbs, M.; Forrer, H. R.; Vogelgsang, S.; Wettstein, F. E.; Schwarzenbach, R. P.; Bucheli, T. D., Occurrence of zearalenone on Fusarium graminearum infected wheat and maize fields in crop organs, soil, and drainage water. *Environ. Sci. Technol.* **2008,** *42*, 5455-5460.
38. Vogelgsang, S.; Sulyok, M.; Hecker, A.; Jenny, E.; Krska, R.; Schuhmacher, R.; Forrer, H. R., Toxigenicity and pathogenicity of *Fusarium poae* and *Fusarium avenaceum* on wheat. *Eur. J. Plant Pathol.* **2008,** *122*, 265-276.
39. Vogelgsang, S.; Jenny, E.; Hecker, A.; Banziger, I.; Forrer, H. R., Fusaria and mycotoxins in wheat - monitoring of harvest samples from growers' fields. *Agrarforschung* **2009,** *16*, 238-243.
40. AWEL Kanton Zürich Baudirektion - Amt für Abfall, Wasser, Energie und Luft (07.02.2012),
41. Logrieco, A.; Mule, G.; Moretti, A.; Bottalico, A., Toxigenic Fusarium species and mycotoxins associated with maize ear rot in Europe. *Eur. J. Plant Pathol.* **2002,** *108*, 597-609.

Supporting information

Figure S6.1: Map of Switzerland including the location of the river water sampling stations: river Glatt catchment (A-E); river Töss catchment (F-I) all from the AWEL monitoring program (▲). NADUF (■) river sampling stations J) Thur; K) Rhine; L) Aare. The agricultural area cropped with winter wheat of the total municipal area is given in percentage (Source: Swiss Federal Statistical Office; Statistic of the agricultural area of Switzerland for the year 2010).

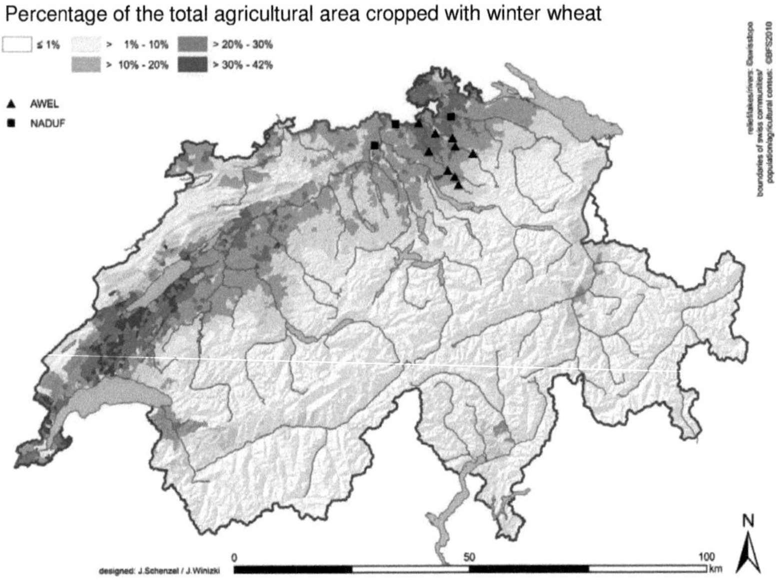

Figure S6.2: Map of the Canton of Zurich and the Canton of Aargau including the location of the river water sampling stations: river Glatt catchment (A-E); river Töss catchment (F-I) all from the AWEL monitoring program (▲). NADUF (■) river sampling stations J) Thur; K) Rhine; L) Aare. The agricultural area cropped with silage maize of the total municipal area is given in percentage (Source: Swiss Federal Statistical Office; Statistic of the agricultural area of Switzerland for the year 2010).

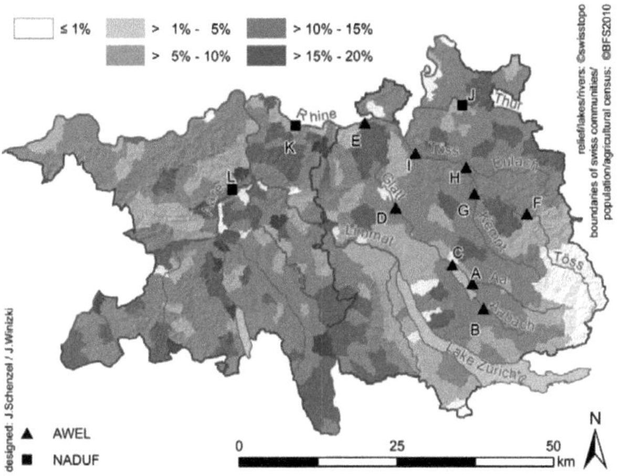

Figure S6.3: Map of the Canton of Zurich and the Canton of Aargau including the location of the river water sampling stations: river Glatt catchment (A-E); river Töss catchment (F-I) all from the AWEL monitoring program (▲). NADUF (■) river sampling stations J) Thur; K) Rhine; L) Aare. The agricultural area cropped with seed corn of the total municipal area is given in percentage (Source: Swiss Federal Statistical Office; Statistic of the agricultural area of Switzerland for the year 2010).

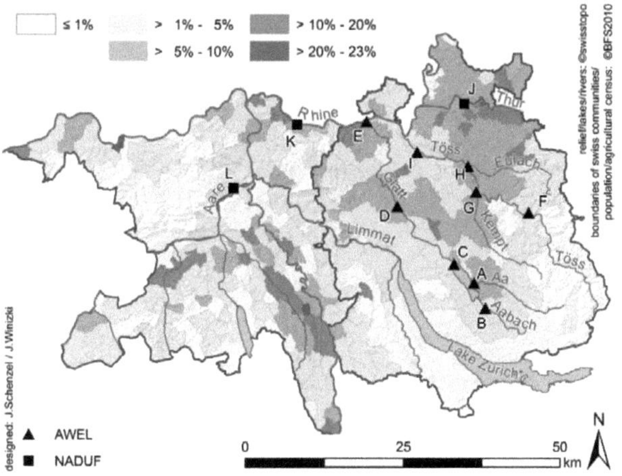

SI-6.2 Analytical procedures and figures of merit.

Raw water samples were filtered (glass fibre filters, pore size 1.2 µm, Millipore, Volketswil, Switzerland) by vacuum filtration (Supelco, Bellfonte PA, USA), transferred to 1 L glass bottles and stored in the dark at +4 °C until analysis within 2 weeks (storage tests showed that mycotoxins were stable over this period of time). Before SPE, the pH was adjusted to between 6.6 and 7.0 by adding either ammonium acetate or acetic acid. In routine analysis the exact volume of 1 L was spiked with 50 µL of the ILIS mixture before the storage or processing of the sample. The samples were shaken vigorously before further treatment. Filtered water samples (1 L) were concentrated and purified by performing reversed-phase SPE (Oasis HLB cartridges, 6 mL, 200 mg; Waters Corporation, Milford MA, USA) on a 12-fold vacuum extraction box (Supelco, Bellfonte PA, USA). The SPE cartridges were consecutively conditioned with 5 mL of MeOH, 5 mL of Milli-Q water and MeOH (1/1, v/v), and 5 mL of Milli-Q water. One Liter water samples were passed through the cartridges with a maximum flow rate of 10 mL/min. Subsequently, the cartridges were washed with 5 mL of Milli-Q water. Without any additional column drying step the analytes were eluted with 5 mL MeOH and the aliquots were collected in conical reaction vial vessels (Supelco, Bellfonte PA, USA). The 5 mL MeOH aliquots were reduced to 100 µL under a gentle nitrogen gas stream at 50°C. The extracts were reconstituted in 900 µL of Milli-Q water/MeOH (90/10, v/v) and transferred into 1.5 mL amber glass vials. The samples were stored at +4°C and analyzed within 48 h.

Chromatographic separation and mass spectrometric detection: LC-MS/MS was performed on a Varian 1200L LC-MS instrument (VarianInc, Walnut Creek CA, USA). The mycotoxins were separated using a 125 mm × 2.0 mm i.d., 3 µm, Pyramid C_{18} column, with a 2.1 mm x 20 mm i.d. 3 µm guard column of the same material (Macherey-Nagel GmbH & Co. KG, Düren, Germany) at room temperature. Two different chromatographic runs were used for the separation of all compounds, on in the positive and one in the negative ionization mode, respectively. The optimized LC mobile phase gradient for the analysis of the analytes measured in negative ionization mode was as follows: 0.0 min: 0% B (100% A), 2.0 min: 0% B, 15.0 min: 100% B, 18.0 min: 100% B, 19.0 min: 0% B, 24.0 min: 0% B. The analytes measured in the positive ionization mode were separated with the following gradient: 0.0 min: 27% B (73% A), 1.0 min: 27% B, 1.3 min: 45% B, 5.0 min: 45% B, 5.3 min: 63% B, 9.0 min: 63% B, 9.3 min: 81% B, 13.0 min: 81% B, 13.3 min: 100% B, 20.0 min: 100% B, 21.0 min: 27% B, 24.0 min: 27% B. In both cases eluent A consisted of Milli-Q water/MeOH/acetic acid (89/10/1, v/v/v) and eluent B of Milli-Q water/MeOH/acetic acid (2/97/1, v/v/v). Both eluents were buffered with 5 mM ammonium acetate. The injection volume was 40 μL and the mobile phase flow was 0.2 mL/min.

LC-MS interface conditions for the ionization of the acidic mycotoxins in the –ESI mode were: needle voltage - 4000 V, nebulizing gas (compressed air) 3.01 bar, drying gas (N_2, 99.5%) 275°C and 1.24 bar, shield voltage -600 V. The neutral mycotoxins were ionized in the +ESI mode: needle voltage + 4500 V, nebulizing gas (compressed air) 3.01 bar, drying gas (N_2, 99.5%) 275°C and 1.24 bar, shield voltage + 600 V.

Table S6.1: Ion suppressions, relative recoveries[a], absolute recoveries [a], and precision of mycotoxins in river water.

Compound	Ion suppression [%]	Concentration level [ng/L]	Relative Recoveries [%]	Absolute Recoveries [%]	Precision [%]
		5	n.a.	n.a.	
3-Acetyl-DON	-10	25	76 (7)	75 (7)	7
		100	87 (7)	77 (6)	
		5	100 (6)	71 (7)	
Deoxynivalenol	-8	25	109 (2)	91 (4)	6
		100	106 (3)	104 (3)	
		5		n.a.	
Nivalenol	39	25		22 (2)	7
		100		21 (2)	
		5		56 (8)	
Beauvericin	-19	25		73 (11)	1
		100		44 (8)	

n.a. not available due to higher method detection limit; [a] Absolute standard deviation of five replicates.

Table S6.2: Ion suppressions, relative recoveries[a], absolute recoveries [a], and precision of mycotoxins in waste water treatment plant effluent.

Compound	Ion suppression [%]	Concentration level [ng/L]	Relative Recoveries [%]	Absolute Recoveries [%]	Precision [%]
		5	n.a.	n.a.	
3-Acetyl-DON	-42	25	104 (7)	79 (7)	9
		100	94 (5)	85 (2)	
		5	106 (12)	33 (8)	
Deoxynivalenol	-61	25	105 (5)	69 (7)	10
		100	96 (6)	72 (3)	
		5		n.a.	
Nivalenol	-28	25		20 (1)	6
		100		11 (0)	
		5		76 (6)	
Beauvericin	18	25		77 (4)	7
		100		104 (8)	

n.a. not available due to higher method detection limit; [a] Absolute standard deviation of five replicates.

Table S6.3: Method detection limit [ng/L][a] of mycotoxins in river water and waste water treatment plant effluent (WWTP).

Compound	MDL river water	MDL WWTP effluent
3-Acetyl-Deoxynivalenol	6.1[b]	6.0[b]
Deoxynivalenol	1.3	1.2
Nivalenol	1.5[b]	1.6[b]
Beauvericin	1.3	3.4

[a] Three times the absolute standard deviation at 5 ng/L; [b] Three times the absolute standard deviation at 25 ng/L.

SI-6.3 Concentrations for all detected mycotoxins [ng/L] at all sampling sites during the investigation period January 2010 till end of October 2011.

Table S6.4: Concentrations of nivalenol [ng/L] in the weekly and fortnightly collected flow-proportional surface water samples from the AWEL and NADUF sampling sites from January 2010 to November 2011.

River location	detected/# samples analyzed	min concentration [ng/L][c]	max concentration [ng/L][c]	median concentration [ng/L][c]	cumulative NIV load [kg][d]	SLF[e] [%]
		River Glatt[a]				
Aa at Niederuster (A)	41/95	1.5[f]	15.8	5.6	0.4	74 ± 7
Aabach at Mönchaltorf (B)	51/95	1.5[f]	20.1	6.4	0.2	74 ± 8
Glatt at Fällanden (C)	21/95	1.5[f]	11.8	5.7	0.4	63 ± 4
Glatt at Oberglatt (D)	51/95	1.5[f]	17.2	5.9	1.1	74 ± 7
Glatt at Rheinsfelden (E)	57/95	1.5[f]	16.8	5.9	1.2	71 ± 7
		River Töss[a]				
Töss at Rämismühle (F)	14/95	1.5[f]	8.8	1.5	0.1	80 ± 2
Kempt at Winterthur (G)	37/95	1.5[f]	24.1	3.4	0.1	72 ± 4
Eulach at Wülflingen (H)	45/95	1.5[f]	15.1	5.2	0.1	83 ± 5
Töss at Freienstein (I)	42/95	1.5[f]	13.8	3.2	0.6	60 ± 4
		Larger Rivers in the Swiss Midlands[b]				
Thur at Andelfingen (J)	23/78	1.5[f]	16.0	6.4	4.7	91 ± 3
Rhine at Rekingen (K)	4/73	1.5[f]	10.5	7.6	13.6	63 ± 1
Aare at Brugg (L)	4/48	1.5[f]	8.0	2.8	2.5	85 ± 1

[a]monitoring network: AWEL (Office for waste, water, energy and air, Canton Zurich, Switzerland); [b]monitoring network: NADUF (National River Monitoring and Survey Program); Letter in parenthesis indicate abbreviations in **Figure 6.1**. [c]out of detected samples. [d]cumulative loads were obtained by a simple addition of all quantified loads which were detected after 22 months. [e] SLF: seasonal load fraction, sum of the NIV loads detected in summer and autumn at each specific station ± uncertainty obtained by an error propagation based on the analytical method precision.[10] [f]above LOD of 1.5 ng/L, but below LOQ of 4.5 ng/L.

Table S6.5: Concentrations of 3-Acetyl-deoxynivalenol [ng/L] in the weekly and fortnightly collected flow-proportional surface water samples from the AWEL and NADUF sampling sites from January 2010 to November 2011.

River location	detected/# samples analyzed	min concentration [ng/L][c]	max concentration [ng/L][c]	median concentration [ng/L][c]	cumulative 3-AcDON load [kg][e]	SLF[e] [%]
		River Glatt[a]				
Aa at Niederuster (A)	41/95	6.1[f]	6.1[f]	6.1[f]	0.06	41 ± 1
Aabach at Mönchaltorf (B)	51/95	6.1[f]	6.1[f]	6.1[f]	0.04	76 ± 2
Glatt at Fällanden (C)	21/95	6.1[f]	6.1[f]	6.1[f]	0.09	42 ± 1
Glatt at Oberglatt (D)	51/95	6.1[f]	19.2	6.1[f]	0.44	49 ± 3
Glatt at Rheinsfelden (E)	57/95	6.1[f]	6.1[f]	6.1[f]	0.86	77 ± 3
		River Töss[a]				
Töss at Rämismühle (F)	14/95	6.1[f]	6.1[f]	6.1[f]	0.03	0
Kempt at Winterthur (G)	37/95	6.1[f]	6.1[f]	6.1[f]	0.10	52 ± 2
Eulach at Wülflingen (H)	45/95	6.1[f]	6.1[f]	6.1[f]	0.03	50 ± 1
Töss at Freienstein (I)	42/95	6.1[f]	6.1[f]	6.1[f]	0.40	46 ± 2
		Larger Rivers in the Swiss Midlands[b]				
Thur at Andelfingen (J)	23/78	6.1[f]	6.1[f]	6.1[f]	0.66	33 ± 1
Rhine at Rekingen (K)	4/73	nd.	nd.	nd.	nd.	-
Aare at Brugg (L)	4/48	6.1[f]	6.1[f]	6.1[f]	0.95	100 ± 1

[a]monitoring network: AWEL (Office for waste, water, energy and air, Canton Zurich, Switzerland); [b]monitoring network: NADUF (National River Monitoring and Survey Program); Letter in parenthesis indicate abbreviations in **Figure 6.1**. [c]out of detected samples. [d]cumulative loads were obtained by a simple addition of all quantified loads which were detected after 22 months. nd.: not detected. [e] SLF: seasonal load fraction, sum of the 3-AcDON loads detected in summer and autumn at each specific station ± uncertainty obtained by an error propagation based on the analytical method precision.[10] [f]above LOD of 6.1 ng/L, but below LOQ of 18.3 ng/L.

Table S6.6: Concentrations of beauvericin [ng/L] in the weekly and fortnightly collected flow-proportional surface water samples from the AWEL and NADUF sampling sites from January 2010 to November 2011.

River location	detected/# samples analyzed	min concentration [ng/L][c]	max concentration [ng/L][c]	median concentration [ng/L][c]	cumulative BEA load [kg][e]	SLF[e] [%]
		River Glatt[a]				
Aa at Niederuster (A)	41/95	1.3[f]	10.5	1.3[f]	0.03	33 ± 1
Aabach at Mönchaltorf (B)	51/95	1.3[f]	22.0	1.3[f]	0.03	23 ± 2
Glatt at Fällanden (C)	21/95	1.3[f]	1.3[f]	1.3[f]	0.03	51 ± 2
Glatt at Oberglatt (D)	51/95	1.3[f]	20.7	1.3[f]	0.13	10 ± 2
Glatt at Rheinsfelden (E)	57/95	1.3[f]	17.3	1.3[f]	0.24	12 ± 2
		River Töss[a]				
Töss at Rämismühle (F)	14/95	1.3[f]	18.0	1.3[f]	0.07	17 ± 2
Kempt at Winterthur (G)	37/95	1.3[f]	7.0	1.3[f]	0.03	25 ± 2
Eulach at Wülflingen (H)	45/95	1.3[f]	5.3	1.3[f]	0.01	8 ± 1
Töss at Freienstein (I)	42/95	1.3[f]	4.8	1.3[f]	0.09	16 ± 2
		Larger Rivers in the Swiss Midlands[b]				
Thur at Andelfingen (J)	23/78	1.3[f]	5.5	1.3[f]	0.22	90 ± 1
Rhine at Rekingen (K)	4/73	nd.	nd.	nd.	nd.	-
Aare at Brugg (L)	4/48	1.3[f]	12.4	4.6	4.39	24 ± 1

[a]monitoring network: AWEL (Office for waste, water, energy and air, Canton Zurich, Switzerland); [b]monitoring network: NADUF (National River Monitoring and Survey Program); Letter in parenthesis indicate abbreviations in **Figure 6.1**. [c]out of detected samples. [d]cumulative loads were obtained by a simple addition of all quantified loads which were detected after 22 months. nd.: not detected. [e] SLF: seasonal load fraction, sum of the BEA loads detected in summer and autumn at each specific station ± uncertainty obtained by an error propagation based on the analytical method precision.[10] [f]above LOD of 1.3 ng/L, but below LOQ of 3.9 ng/L.

SI-6.5 Concentration versus discharge: dilution as concentration determinant.

Figure S4: Mycotoxin concentrations depicted versus the average weekly river discharge [m^3/s] in the rivers Glatt (●), Töss (●), and the three sampled NADUF (▽) rivers (Thur, Rhine and Aare) showing the dilution as one of the main determinant. **A)** Deoxynivalenol, **B)** Nivalenol, **C)** 3-Acetyl-deoxynivalenol, and **D)** Beauvericin. LOD = limit of detection; LOQ = limit of quantification.

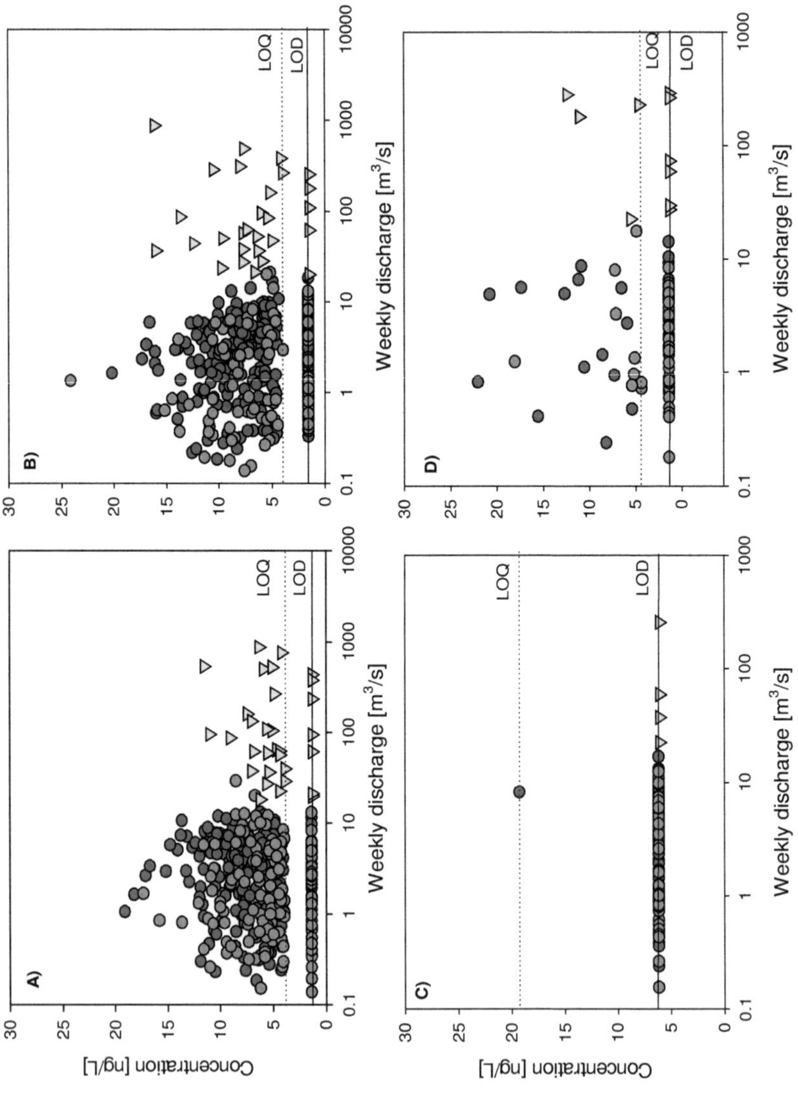

SI-6.6 Occurrence of mycotoxins in all sampled river water stations from January 2010 until November 2011.

Figure S6.5: Occurrence of mycotoxins [ng/L] in river water samples from AWEL station Aa at Niederuster (Station A). Left axis (A) 3-acetyl-deoxynivalenol [ng/L]; (B) nivalenol [ng/L]; (C) deoxynivalenol [ng/L]; (D) beauvericin [ng/L]; (E) weekly river water discharge [m^3/s]. The colored stripes indicate the seasons: spring: green; summer: red; autumn: yellow; winter: white. LOD: limit of detection; points under LOD-level: not detected. Note that on the right axis of the graphs for panel A-D the corresponding loads [g] are given.

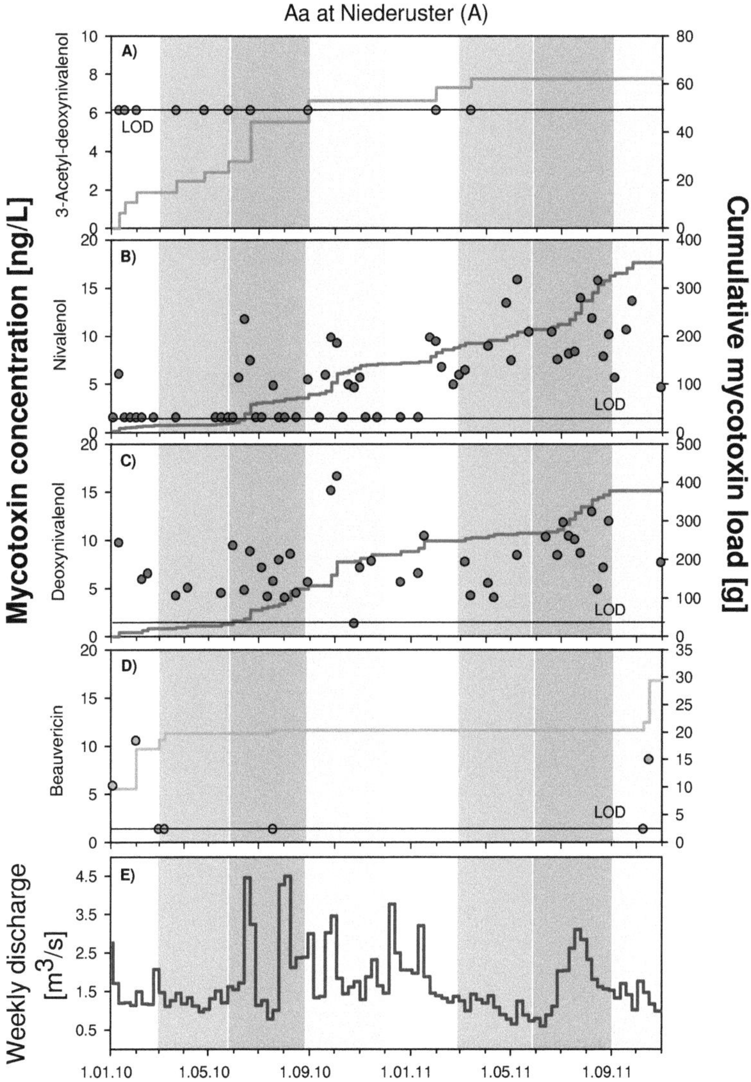

Figure S6.6: Occurrence of mycotoxins in river water samples from AWEL station Aabach at Mönchaltorf (Station B). Left axis (A) 3-acetyl-deoxynivalenol [ng/L]; (B) nivalenol [ng/L]; (C) deoxynivalenol [ng/L]; (D) beauvericin [ng/L]; (E) weekly river water discharge [m^3/s]. The colored stripes indicate the seasons: spring: green; summer: red; autumn: yellow; winter: white. LOD: limit of detection; points under LOD-level: not detected. Note that on the right axis of the graphs for panel A-D the corresponding loads [g] are given.

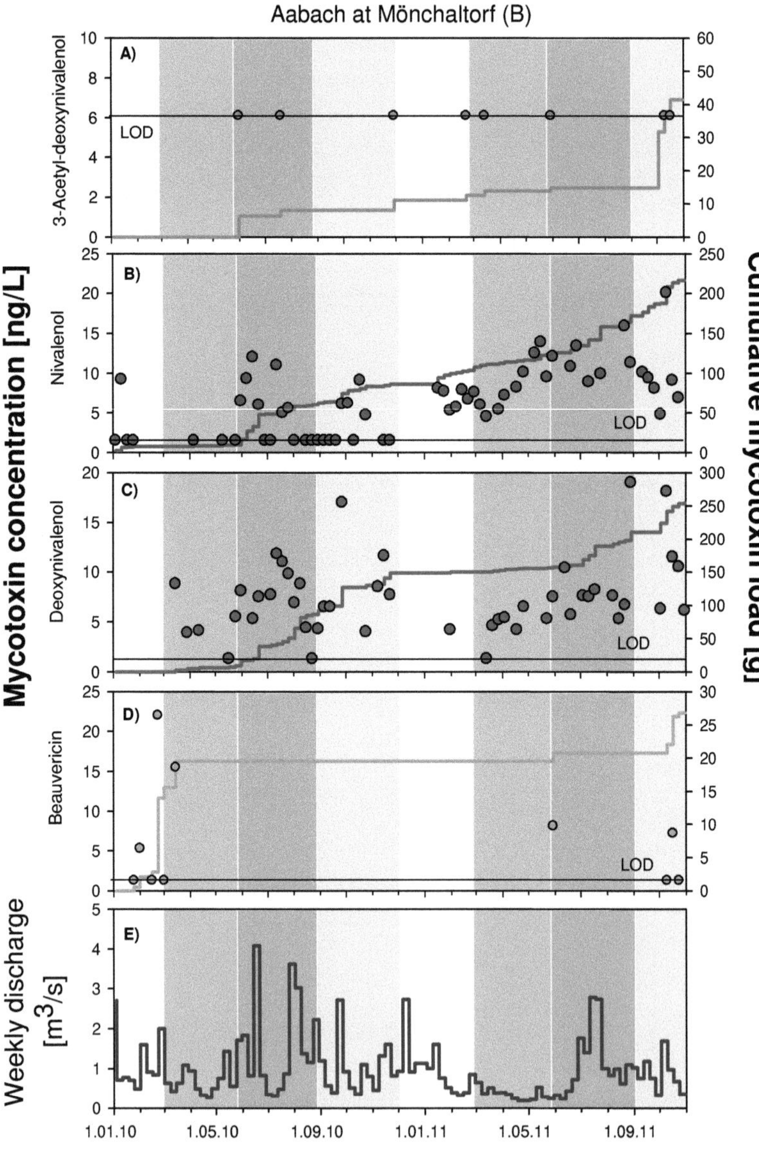

Figure S6.7: Occurrence of mycotoxins in river water samples from AWEL station Glatt at Fällanden (Station C). Left axis (A) 3-acetyl-deoxynivalenol [ng/L]; (B) nivalenol [ng/L]; (C) deoxynivalenol [ng/L]; (D) beauvericin [ng/L]; (E) weekly river water discharge [m^3/s]. The colored stripes indicate the seasons: spring: green; summer: red; autumn: yellow; winter: white. LOD: limit of detection; points under LOD-level: not detected. Note that on the right axis of the graphs for panel A-D the corresponding loads [g] are given.

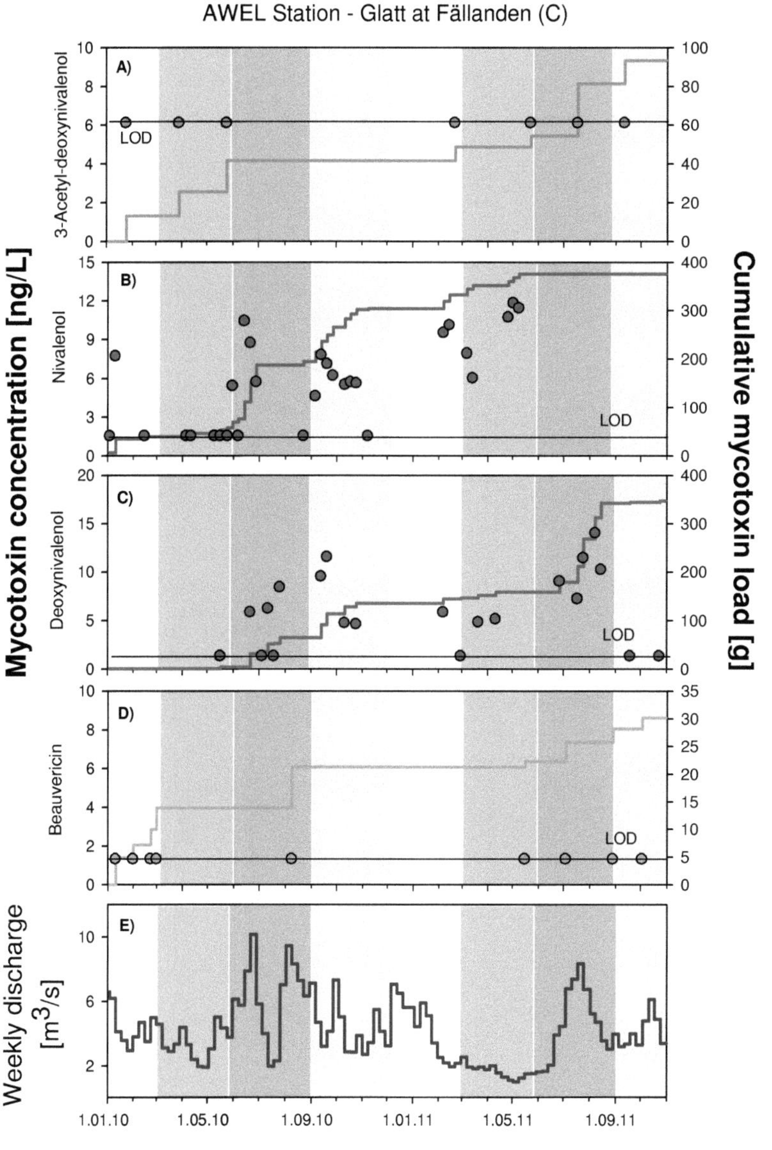

Figure S6.8: Occurrence of mycotoxins in river water samples from AWEL station Glatt at Rheinsfelden (Station E). Left axis (A) 3-acetyl-deoxynivalenol [ng/L]; (B) nivalenol [ng/L]; (C) deoxynivalenol [ng/L]; (D) beauvericin [ng/L]; (E) weekly river water discharge [m^3/s]. The colored stripes indicate the seasons: spring: green; summer: red; autumn: yellow; winter: white. LOD: limit of detection; points under LOD-level: not detected. Note that on the right axis of the graphs for panel A-D the corresponding loads [g] are given.

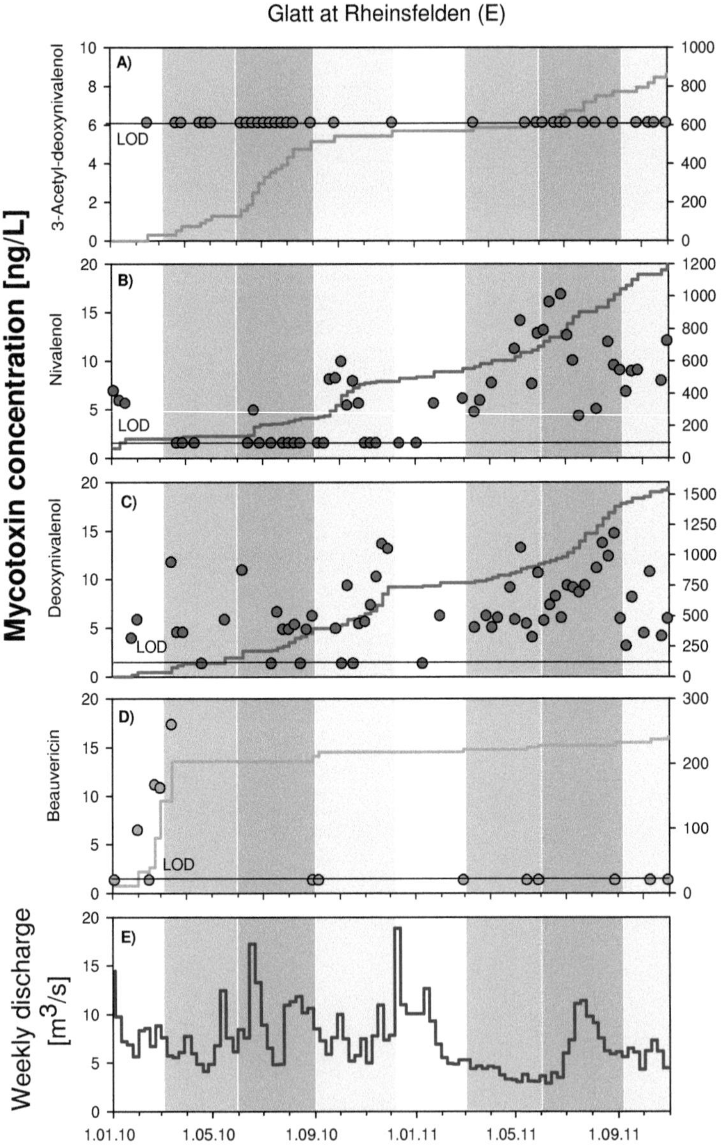

Figure S6.9: Occurrence of mycotoxins in river water samples from AWEL station Töss at Rämismühle (Station F): Left axis (A) 3-acetyl-deoxynivalenol [ng/L]; (B) nivalenol [ng/L]; (C) deoxynivalenol [ng/L]; (D) beauvericin [ng/L]; (E) weekly river water discharge [m^3/s]. The colored stripes indicate the seasons: spring: green; summer: red; autumn: yellow; winter: white. LOD: limit of detection; points under LOD-level: not detected Note that on the right axis of the graphs for panel A-D the corresponding loads [g] are given.

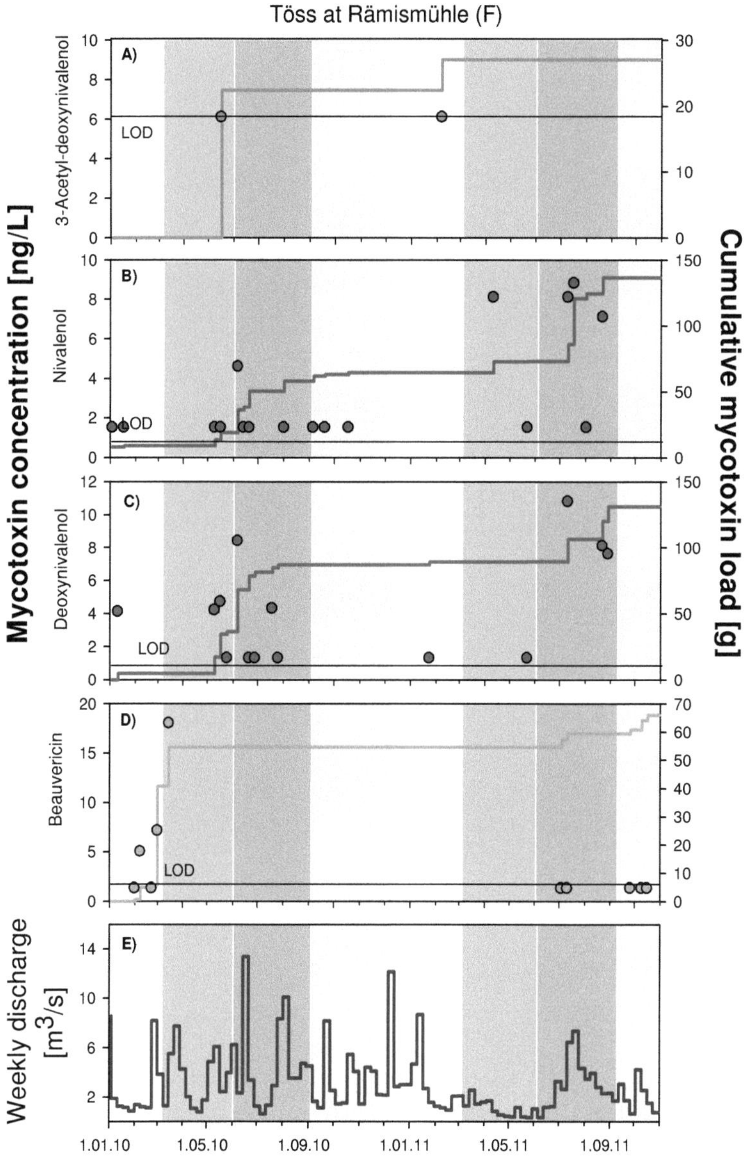

Figure S6.10: Occurrence of mycotoxins in river water samples from AWEL station Kempt at Winterthur (Station G): Left axis (A) 3-acetyl-deoxynivalenol [ng/L]; (B) nivalenol [ng/L]; (C) deoxynivalenol [ng/L]; (D) beauvericin [ng/L]; (E) weekly river water discharge [m^3/s]. The colored stripes indicate the seasons: spring: green; summer: red; autumn: yellow; winter: white. LOD: limit of detection; points under LOD-level: not detected. Note that on the right axis of the graphs for panel A-D the corresponding loads [g] are given.

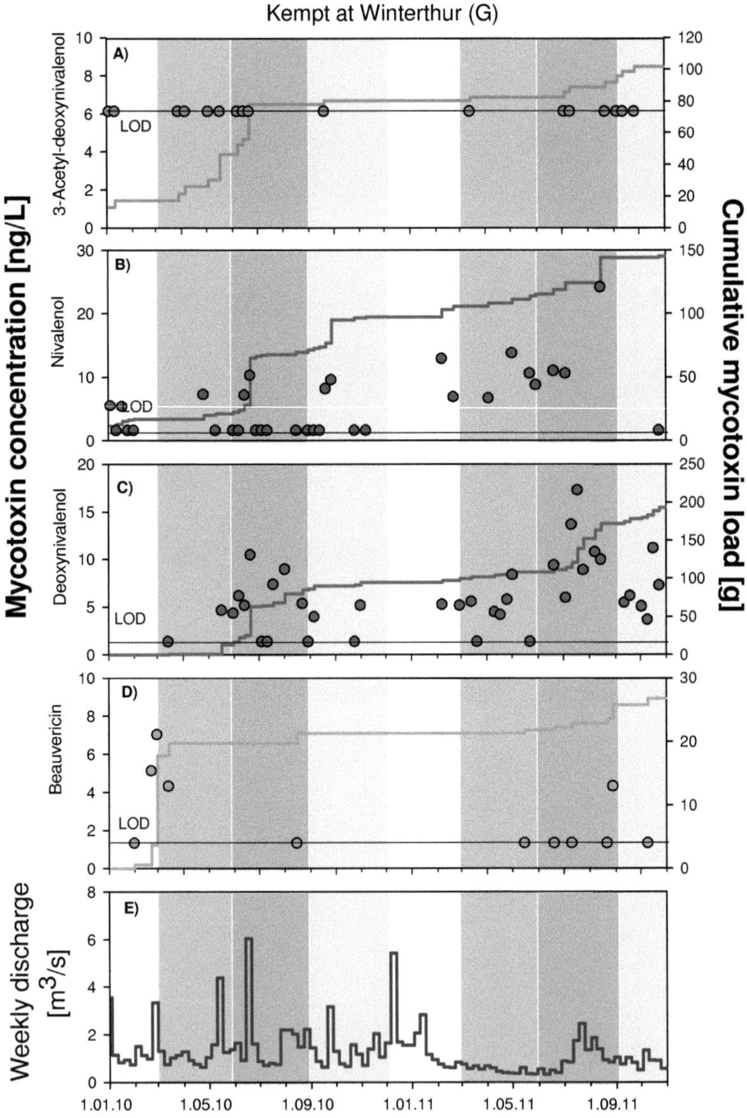

Figure S6.11: Occurrence of mycotoxins in river water samples from AWEL station Eulach at Wülfingen (Station H): Left axis (A) 3-acetyl-deoxynivalenol [ng/L]; (B) nivalenol [ng/L]; (C) deoxynivalenol [ng/L]; (D) beauvericin [ng/L]; (E) weekly river water discharge [m^3/s]. The colored stripes indicate the seasons: spring: green; summer: red; autumn: yellow; winter: white. LOD: limit of detection; points under LOD-level: not detected. Note that on the right axis of the graphs for panel A-D the corresponding loads [g] are given.

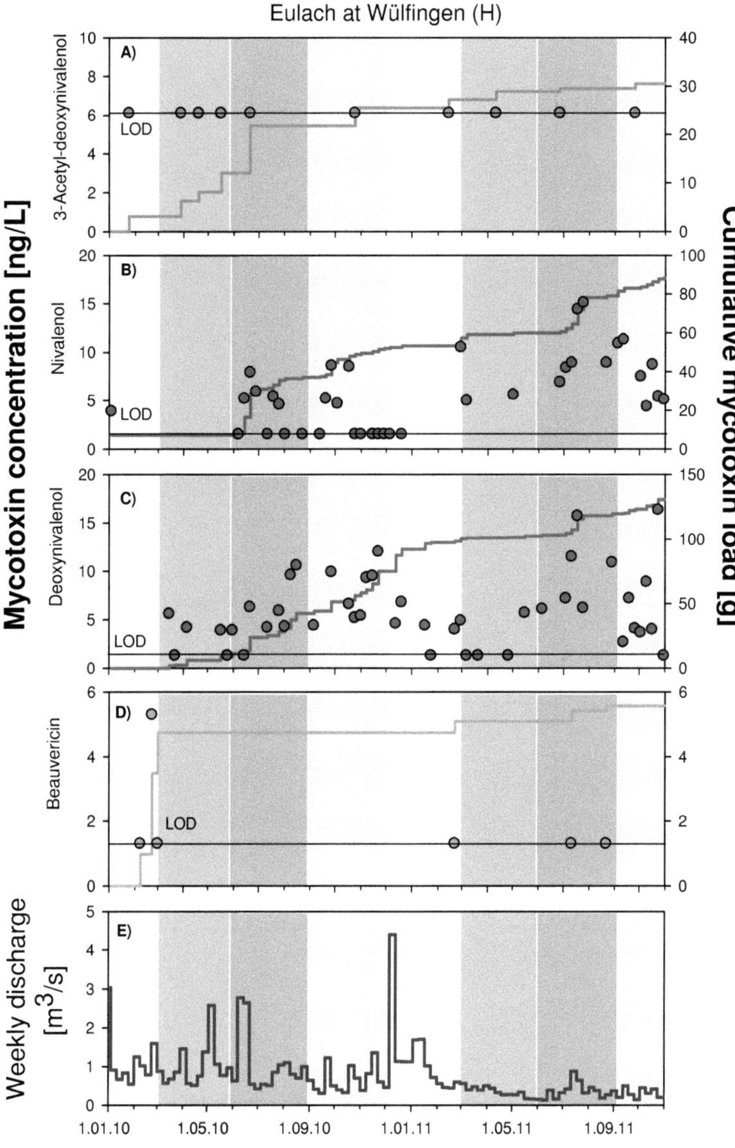

Figure S6.12: Occurrence of mycotoxins in river water samples from AWEL station Töss at Freienstein (Station I): Left axis (A) 3-acetyl-deoxynivalenol [ng/L]; (B) nivalenol [ng/L]; (C) deoxynivalenol [ng/L]; (D) beauvericin [ng/L]; (E) weekly river water discharge [m^3/s]. The colored stripes indicate the seasons: spring: green; summer: red; autumn: yellow; winter: white. LOD: limit of detection; points under LOD-level: not detected. Note that on the right axis of the graphs for panel A-D the corresponding loads [g] are given.

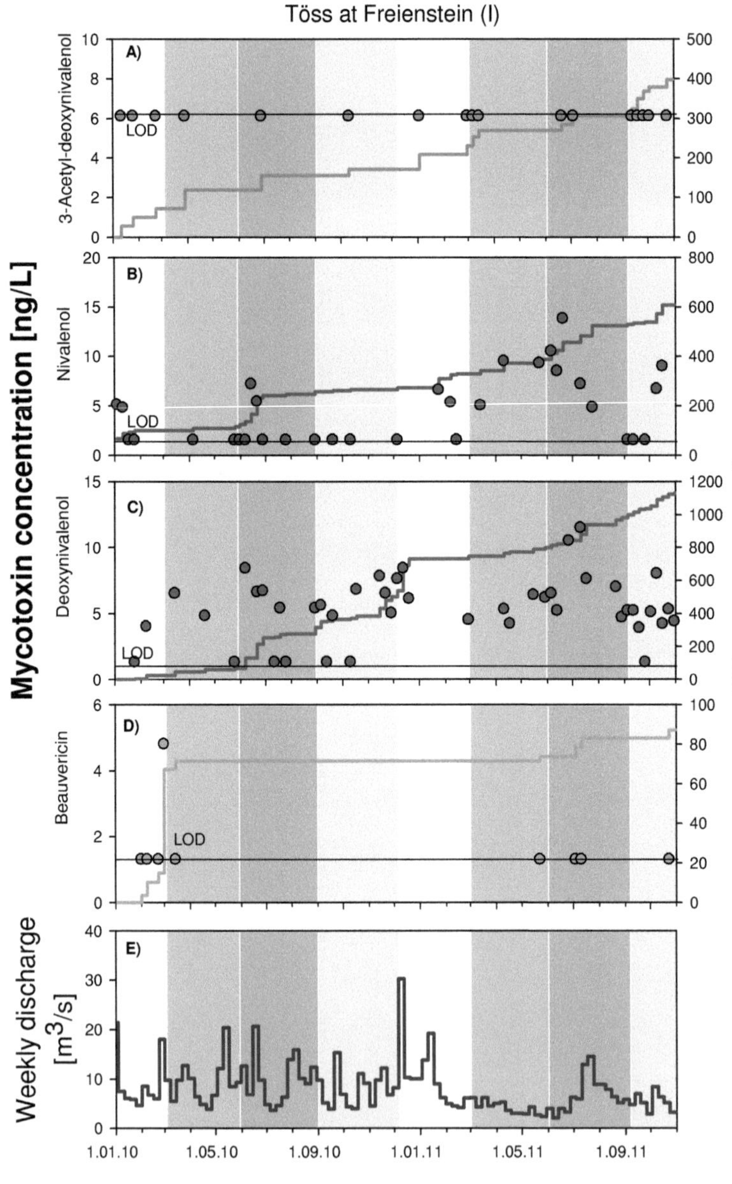

SI-6.7 Seasonal load fractions of mycotoxins.

Figure S6.13: Seasonal load fraction (SLF) of each mycotoxin versus the ratio of total cropped wheat area and number of inhabitants. **A**) Deoxynivalenol, **B**) Nivalenol, **C**) 3-Acetyl-deoxynivalenol, and **D**) Beauvericin.

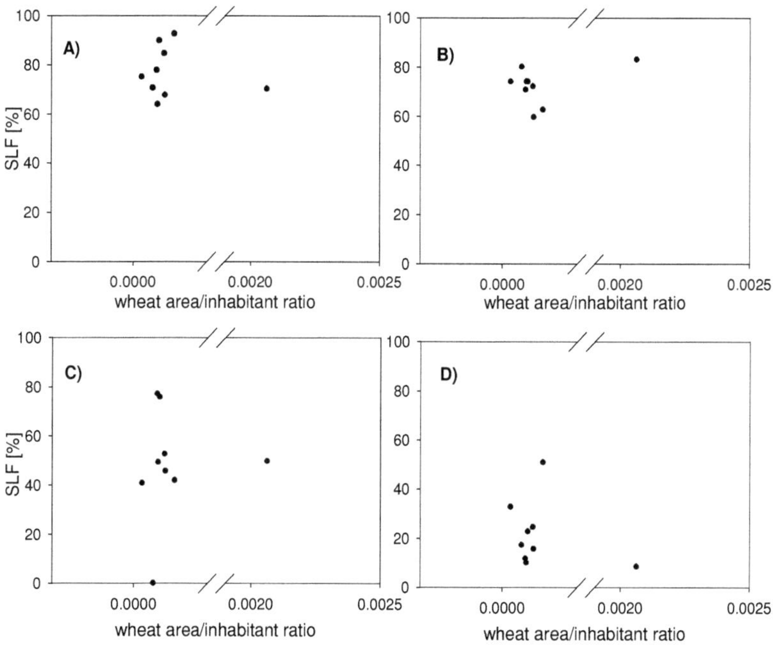

TABLE S6.7: Linear regression analysis of cumulative mycotoxin loads over the period of nearly two years of investigation and during summertime 2010 vs. total cropped wheat area and population equivalents in the corresponding river catchment. The cumulative mycotoxin loads are at log scale.[a]

	vs. wheat area					vs. inhabitants				
compound	R^2	MSE	p-value	slope	intercept	R^2	MSE	p-value	slope	intercept
January 2010 - November 2011:										
Deoxynivalenol	0.46	0.000854	0.044	15.23	-0.75	0.89	4.19E-07	0.00014	215084.2	-0.85
Nivalenol	0.31	0.00171	0.119	13.72	-0.72	0.91	8.37E-07	0.00008	238745.6	-0.87
3-Acetyl-DON	0.54	0.00589	0.025	13.3	-1.4	0.91	5.78E-07	0.00007	176301.9	-1.3
Beauvericin	0.17	0.000553	0.271	8.56	-1.56	0.65	2.71E-07	0.00897	170201.4	-1.74
June 2010 – October 2010:										
Deoxynivalenol	0.5	0.00293	0.033	18.03	-1.08	0.87	2.88E-07	0.00025	241809.2	-1.15
Nivalenol	0.19	0.00301	0.248	11.71	-1.08	0.82	2.95E-07	0.00076	251366.7	-1.24
3-Acetyl-DON	0.31	0.00457	0.193	9.3	-1.8	0.83	4.49E-07	0.00443	158626.6	-2
Beauvericin	0.17	0.000217	0.271	8.56	-1.56	0.65	2.13E-08	0.00897	170201.4	-1.74

[a]to obtain normally distributed values. Note that there is no correlation between wheat area and population equivalent (R^2=0.41; p-value=0.35204). The Tukey-Anscombe plot showed normal distributed residuals. MSE = mean squared error.

SI-6.8 Estimated mycotoxin fractions emitted from agricultural areas cropped with winter wheat and from human excretion via WWTP effluents over an investigation period from January 2010 until November 2011 in the river Glatt and Töss catchment. The width of green and blue boxes represent winter wheat area per catchment [km^2] and average water flow Q [m^3/s] as observed during the period of investigation, respectively.

Table S6.8: Predicted loads [g] over the whole investigation period of mycotoxins coming from human emission via WWTP and from agricultural areas cropped with small grain cereals.

total loads [g]	**WWTP**				**agriculture**			
	DON	NIV	3-Ac-DON	BEA	DON	NIV	3-Ac-DON	BEA
Glatt catchment								
Aabach at Mönchaltorf (A)	123 ± 47	32 ± 39	51 ± 27	9 ± 1	177 ± 309	106 ±	30 ± 27	-
Aa at Niederuster (B)	587 ±	154 ±	242 ± 128	14 ± 6	282 ± 491	141 ±	21 ± 19	-
Glatt at Fällanden (C)	251 ± 96	66 ± 81	104 ± 55	18 ± 3	0.6 ± 0.9	0.4 ± 0.6	0.05 ± 0.04	-
Glatt at Oberglatt (D)	839 ±	220 ±	346 ± 183	60 ± 8	567 ± 986	396 ±	84 ± 75	-
Glatt at Rheinsfelden (E)	1366 ±	358 ±	564 ± 299	98 ± 14	1735 ±	720 ±	154 ± 138	-
Töss catchment								
Töss at Rämismühle (F)	79 ± 30	21 ± 26	33 ± 17	6 ± 1	84 ± 146	79 ± 122	9 ± 8	-
Kempt at Winterthur (G)	202 ± 77	53 ± 65	83 ± 44	14 ± 2	168 ± 293	144 ±	17 ± 15	-
Eulach at Wülfingen (H)	53 ± 20	14 ± 17	22 ± 12	4 ± 1	409 ± 711	170 ±	44 ± 39	-
Töss at Freienstein (I)	1505 ±	394 ±	621 ± 329	108 ± 15	1231 ±	585 ±	231 ± 207	-

Figure S6.14: Surface waters of the Canton of Zurich with the river Glatt and Töss highlighted in dark blue. Inserted bar charts show the estimated amounts from WWTP effluents [g] (blue), and from agricultural areas cropped with winter wheat [g] (green). They are compared to the total nivalenol load [g] (red) at each sampling station over an investigation period from January 2010 until November 2011.

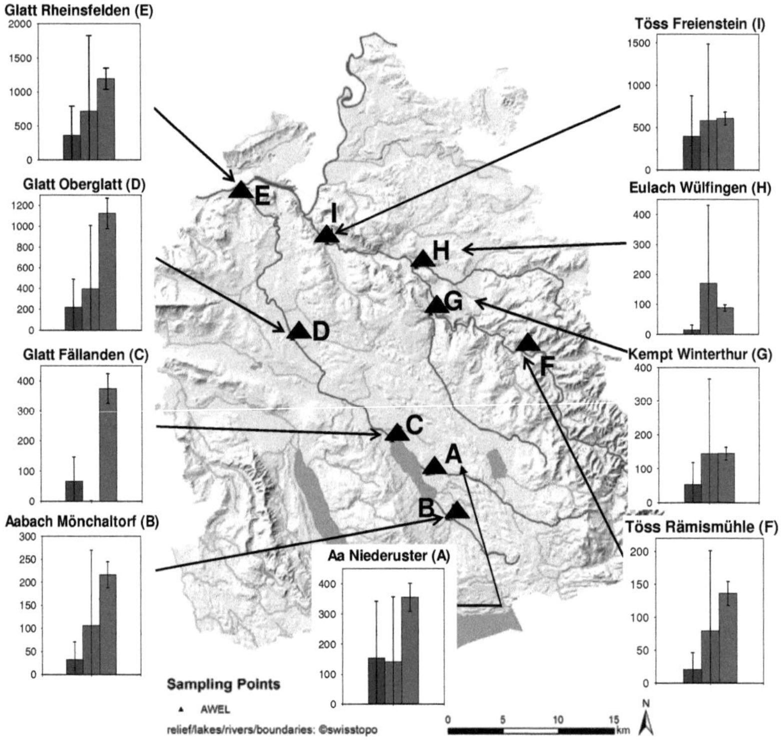

Figure S6.15: Surface waters of the Canton of Zurich with the river Glatt and Töss highlighted in dark blue.Inserted bar charts show the estimated amounts from WWTP effluents [g] (blue), and from agricultural areas cropped with winter wheat [g] (green). They are compared to the total 3-Acetyl-deoxynivalenol load [g] (red) at each sampling station over an investigation period from January 2010 until November 2011.

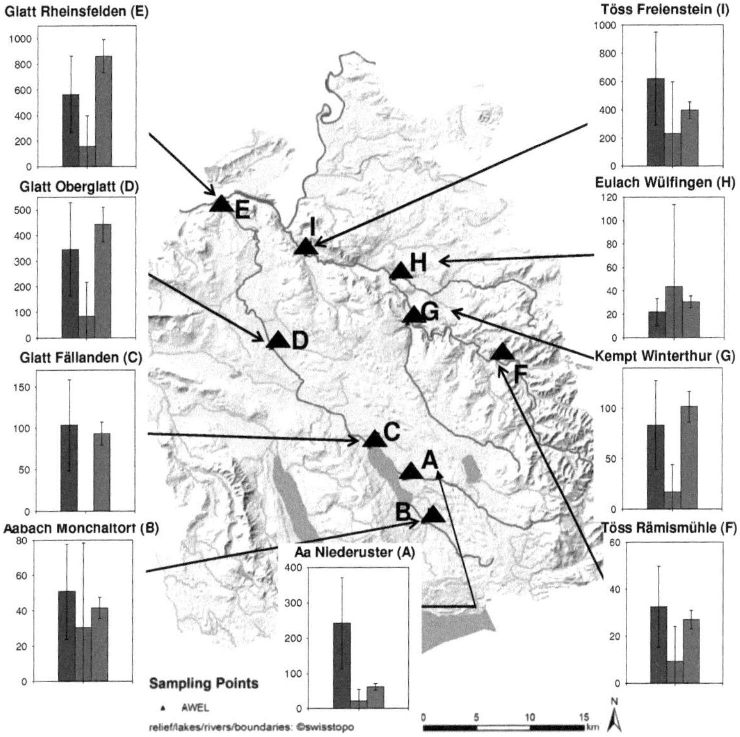

Figure S6.16: Surface waters of the Canton of Zurich with the river Glatt and Töss highlighted in dark blue. Inserted bar charts show the estimated amounts from WWTP effluents [g] (blue), and from agricultural areas cropped with winter wheat [g] (green). They are compared to the total beauvericin load [g] (red) at each sampling station over an investigation period from January 2010 until November 2011.

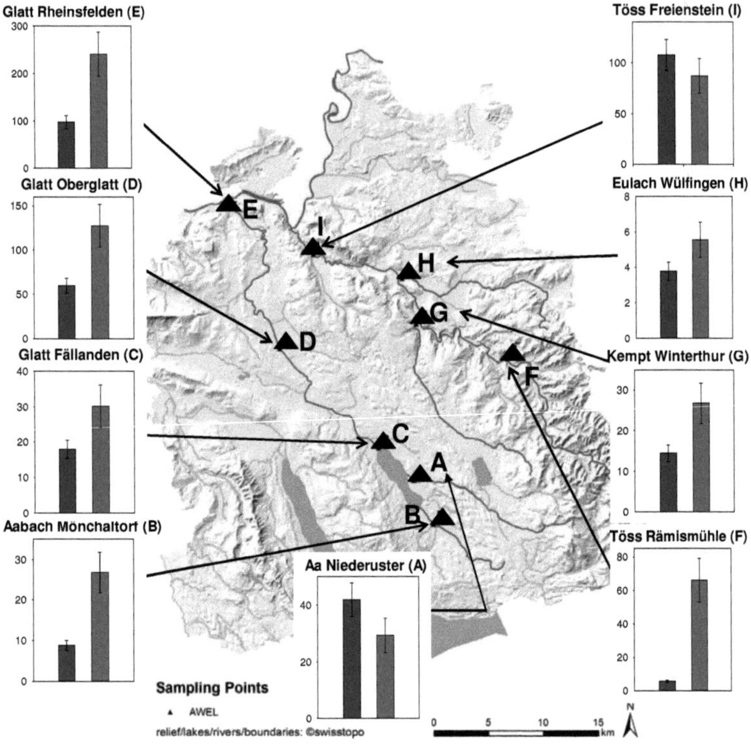

CHAPTER 7:

CONCLUSIONS AND OUTLOOK

7.1 Conclusions

Based on the objectives of this doctorial thesis four different hypotheses were formulated in chapter 1, which are discussed conclusively according to the obtained results.

I. Mycotoxins are emitted into the (aqueous) environment from agricultural fields or via waste water treatment plants (WWTP):

To be able to accept or reject this first hypothesis, two analytical methods were developed to quantify a wide range of mycotoxins in whole wheat plants, and in three different aqueous matrices, respectively. These methods proved to be suitable to quantify the selected mycotoxins at the required low $mg/kg_{dw.}$, and ng/L range, respectively, and with the desired accuracy. Especially, the use of six isotope-labeled, and one deuterated internal standards for the analysis of mycotoxins in aqueous samples facilitated the correction of matrix related problems during sample preparation and ionization of the analytes in the LC-MS interface. For those compounds where no isotope-labeled analogues were available, matrix-matched calibrations were employed.

The analysis of mycotoxins in whole wheat plants during two years indicated that mycotoxin concentrations and patterns in wheat plants varied significantly (by a factor of max. 600) both over time and in between the different artificially infected subfields. For several of the detected mycotoxins in whole wheat plants the concentrations reached a maximum level already two to three weeks before harvest, and declined thereafter. Breakdown-, and conjugation processes, as well as the wash-off of water soluble mycotoxins from the plants are plausible explanations for the prior-harvest decline. The considerable variations of mycotoxin amounts in whole wheat plants reflect the natural variability, e.g., caused by variable weather conditions and the different artificially applied *Fusarium* species. Although only a minor fraction of the total amounts produced in the wheat plants is emitted into the drainage water during rain events, this emission is still relevant for the occurrence of mycotoxins in river water. The study revealed that the main mycotoxin fraction is removed from the agricultural field with the harvest, but as previously demonstrated, within the growing season the emission via drainage water is an important input source of mycotoxins in the aquatic environment.

Additionally, four out of the five most prominently detected mycotoxins in drainage water samples were also regularly found in a Swiss WWTP effluent, demonstrating that human excretion is another possible and important emission source of mycotoxins in surface waters. Interestingly, only field produced mycotoxins prevailed in the WWTP effluent samples.

Finally, the analysis of mycotoxins in river water samples revealed that the same compounds were regularly present in Swiss surface waters. Therefore, the first hypothesis can be fully accepted.

II. The relative importance of these input pathways depends on the mycotoxin type, and the produced amount, and whether they are produced in the field or during storage:

The two main input pathways were identified and already discussed above. Generally, field and storage derived mycotoxins can be present in food products. Accordingly, field and storage derived mycotoxins should have been detectable in the different aqueous samples. The fact, that only field produced mycotoxins were detected in all aqueous matrices indicated that a certain minimum amount of mycotoxins need to be produced by the fungi to be detectable in the environment. Consequently, all mycotoxins produced during storage are absent in this study. This can be explained by the fact that they are produced in much smaller amounts, and hence their potential total flux derived from WWTP effluents is much smaller due to concentrations, which were always below the LOQ.

Besides this, we expected that the relative importance of *Fusarium* infected crops and WWTP as sources of field-produced mycotoxins to be higher and lower, respectively, in rural, and *vice versa* in urban regions. However, this could not be demonstrated for any of the mycotoxins investigated and might be explained by the fact that there are other important possible sources of mycotoxins, e.g., emissions from husbandry animal excretions, for which, so far no reliable numbers are available.

Nevertheless, according to the above presented results, the second hypothesis can be accepted.

III. The magnitude of emission depends on the produced amount, persistence, and chemical-physical properties of the individual mycotoxin:

As pointed out along with the hypothesis II, the magnitude of emission strongly depends on the produced amount of mycotoxins. Additionally, the chemical-physical properties of each compound influence the persistence and mobility, hence to which extent the mycotoxin is emitted. To date, experimentally determined data for mycotoxins are scarce. Accordingly, a dynamic HPLC-based flow-through method was applied to obtain sorption coefficients to soil organic matter (K_{oc}) for over 30 mycotoxins. This experimental data set is of great value for risk assessment of mycotoxins, because so far, virtually no data were available for this type of compounds, and, as also demonstrated, presently available methods for prediction of K_{oc} values fail, mostly because of the complex molecular structure of the compounds. Besides, the attained partitioning coefficients help to describe at least approximately the transport of mycotoxins through the different compartments (soil column, and different stages of a WWTP) into the aquatic environment. Further, according to a simple rule of thumb a compound with a Log $K_{oc} >> 3$ is assumed to be removed from the aqueous/mobile phase and will therefore not show up to a significant extent in surface waters. Even though the presence of beauvericin in all different aqueous matrices was somewhat surprising, the third hypothesis can be accepted as well.

IV. Certain mycotoxins are of relevance in terms of their environmental risk to aquatic ecosystems:

Based on the quantified mycotoxin concentrations in drainage and river waters, as well as the identified contributions from the two main sources, the mycotoxin exposure in surface waters is rather low. Until now, only a few studies are available dealing with the ecotoxicity of these compounds.[1-5] Yet, the ecotoxicological relevance of the low levels of quantified mycotoxins seems to be marginal for relative dynamic water bodies, like streams, rivers, and lakes, as proposed by Maragos et al.[6] Still, further studies are needed to accept or reject this hypothesis in absolute terms.

7.2 Outlook

Although the river water monitoring programs used in this work are largely representative for Swiss midland rivers, there may be occasions/locations that would merit further investigations, e.g. in areas of intense crop production, or in regions where different crops are grown (for example maize in the canton of Ticino, see[7]).

Additionally, another research question arises regarding the husbandry animal excretion input source, because a complete and explicit source apportionment was not possible so far. Husbandry animal excretions as a third emission source could not be included, due to the fact that no complete data set was available. Therefore the mycotoxin content of manure needs to be determined with a suitable analytical method. Currently, just a few analytical methods are available, dealing with this kind of matrix,[8, 9] but further mycotoxins need to be included to obtain a broad insight of the mycotoxin content in manure samples.

Overall, the environmentally relevant compartments soil and manure are still not investigated systematically for the presence of mycotoxins. Here, the transport processes of mycotoxins from the plant through the soil column are not fully understood. Thus, the transport through the soil column could be investigated in more detail either by lysimeter studies or with lab soil column experiments. The advantages of lysimeter studies would be the sampling of percolating water at different soil layers. Especially interesting in these terms would be the impact of changing chemical, and/or physical conditions, like pH, temperature, and oxic-anoxic conditions, which might influence their stability, mobility and bioavailability. The changing parameters could be tested with soil column experiments.

Additionally, the fate of mycotoxins in WWTP effluents was studied, and especially for DON a poor elimination rate was reported.[10] According to the omnipresence of DON in all WWTP compartments, the use of further treatment processes like ozonation or activated carbon filtration could increase the elimination rate substantially. So, further sorption studies with activated carbon could be a useful first indicator, if mycotoxins, as another class of aquatic micropollutants, can be eliminated during an enhanced waste water treatment process.

Aside from mycotoxins, the existence of about 1500 other natural toxins is documented.[11] The here presented study about the environmental fate and behavior of mycotoxins is only the tip of the iceberg. It is neither possible, nor reasonable to conduct such intensive studies for all natural toxins. According to the outcome of this study, we would propose a rather systematical approach. Conducting an extensive literature search for chemical-physical data points (e.g., K_{ow}, K_{oc}), and measured concentrations in different matrices on each compound should be used as a first rational indicator for their potential environmental presence. Additionally, information about the toxicity of these compounds would be useful to integrate. Consequently, the number of potential environmentally relevant natural toxins decreases drastically, and it could be further on decided for which compounds a more intensive study would be useful.

References (chapter 7)

1. Schwartz, P., et al., Life-cycle exposure to the estrogenic mycotoxin zearalenone affects zebrafish (Danio rerio) development and reproduction. *Environmental Toxicology* in press.
2. Schwartz, P., et al., Short-term exposure to the environmentally relevant estrogenic mycotoxin zearalenone impairs reproduction in fish. *Science of the Total Environment* **2010,** *409*, 326-333.
3. Fornelli, F., et al., Cytotoxicity induced by nivalenol, deoxynivalenol, and fumonisin B, in the SF-9 insect cell line. *In Vitro Cell. Dev. Biol.-Anim.* **2004,** *40*, 166-171.
4. Fotso, J.; Smith, J. S., Evaluation of beauvericin toxicity with the bacterial bioluminescence assay and the Ames mutagenicity bioassay. *J. Food Sci.* **2003,** *68*, 1938-1941.
5. Sarter, S., et al., Effects of mycotoxins, aflatoxin B(1) and deoxynivalenol, on the bioluminescence of Vibrio fischeri. *World Mycotoxin J.* **2008,** *1*, 189-193.
6. Maragos, C. M., Zearalenone occurrence in surface waters in central Illinois, USA. *Food Additives and Contaminants: Part B* **2012,** *5*, 55-64.
7. Dorn, B., et al., Fusarium species complex on maize in Switzerland: occurrence, prevalence, impact and mycotoxins in commercial hybrids under natural infection. *Eur. J. Plant Pathol.* **2009,** *125*, 51-61.
8. Hartmann, N., et al., Quantification of zearalenone in various solid agroenvironmental samples using D-6-zearalenone as the internal standard. *J. Agric. Food Chem.* **2008,** *56*, 2926-2932.
9. Schollenberger, M., et al., Simultaneous determination of a spectrum of trichothecene toxins out of residuals of biogas production. *J. Chromatogr. A* **2008,** *1193*, 92-96.
10. Wettstein, F. E.; Bucheli, T. D., Poor elimination rates in waste water treatment plants lead to continuous emission of deoxynivalenol into the aquatic environment. *Wat. Res.* **2010,** *44*, 4137-4142.
11. Harborne, J. B., Recent advances in chemical ecology. *Nat. Prod. Rep.* **1997,** *14*, 83-98.

DANKSAGUNG

Ganz herzlich möchte ich mich bei meinen beiden Doktorvätern Prof. Dr. René P. Schwarzenbach und Prof. Dr. Konrad Hungerbühler dafür bedanken, dass sie meine Dissertation mit fachkundigem Wissen betreut haben. Danke Konrad, dass Du meine Dissertation als Doktorvater so kurzfristig am Ende noch als Hauptbetreuer übernommen hast und bei den Korrekturen der Publikationen immer noch den einen oder anderen „Bock" entdeckt hast.

Ein besonderer Dank gilt meinem Betreuer Dr. Thomas D. Bucheli, der diese vielseitige Doktorarbeit im Rahmen eines BAFU-finanzierten Projektes erst ermöglicht hat. Ein grosser Dank vor allem für Deine zeitnahen, konstruktiven und förderlichen Korrekturen zu Manuskripten, Abstracts, Vorträgen und Postern. Danke für die vielen fachlichen Diskussionen, die meine Arbeit und umweltchemische Sichtweise nachhaltig bereichert haben.

Prof. Dr. Thorsten Reemtsma danke ich für die Bereitwilligkeit als externer Co-Referent meine Doktorarbeit zu begutachten.

Besonders möchte ich mich bei Steven Droge bedanken, für die sehr lehrreiche und spannende Zusammenarbeit im Bereich der Sorption von Mykotoxinen. Mein weiterer Dank richtet sich an die gesamte Gruppe der Analytischen Umweltchemie am UFZ Leipzig von Prof. Dr. Kai-Uwe Goss. Danke für Eure schnelle Integration in die Gruppe und die schöne gemeinsame Weihnachtsfeier.

Einen grossen Dank richte ich an Dr. Hans Jörg Bachmann für die zusätzliche finanzielle Unterstützung meiner Doktorarbeit, insbesondere im Bereich von benötigten Labormaterialen und sehr teuren Isotopenmarkierten internen Standards.

Die tatkräftige Unterstützung im Bereich der Probenaufarbeitung für die Analytik von Mykotoxinen in wässrigen Matrices habe ich Pakeerathan Srikanthan und Stefan Weissen zu verdanken.

Für das angenehme Arbeitsklima und die erheiternden Diskussionen danke ich meinen jetzigen und früheren Labor- und BürokollegInnen, insbesondere Tripti Agarwal, Fränzi Blum, Louise Camenzouli, Xanat Flores-Cervantes, Corinne Hörger, Lungile Lukele, Fabienne Schwab, Anna Sobek, und Asma Younas. Besonders habe ich das gegenseitige Aushelfen im Labor und auf dem Feld geschätzt.

Ebenfalls gilt mein Dank dem gesamten Analytik-Supportbereich der ART, insbesondere Hans-Ruedi Bosshard, Jolanda Bucher, Diane Bürge, Erika Hinnen, Agush Loshi, Johanna Rauwolf, Barbara Ropka, Hans Stünzi, Brigitte Weiss und Sabina Wojdyla.

Gezielt möchte ich mich bei verschiedenen Forschungsbereichsgruppen der ART bedanken, die zum Gelingen meiner Doktorarbeit beigetragen haben; allen voran der Feldgruppe und der Gruppe Ökologischer Pflanzenschutz von Susanne Vogelgsang und Hans-Ruedi Forrer.

Ein grosser Dank geht von ganzem Herzen an all meine Freundinnen und Freunde, sowie an den erweiterten Familienkreis. Danke für all eure Unterstützung und Zuwendung, die offenen Ohren, die hilfreichen Ratschläge und die Diskussionen, die ich während meiner Dissertationszeit von euch erfahren habe. Ohne euch wäre ich nicht da, wo ich heute bin!

Ein liebevoller Dank richtet sich insbesondere an Gerd Rothardt. Danke für Deinen bewundernswerten, rationalen Blick auf Probleme und die immer währende Unterstützung in allen Bereichen. Danke, dass Du mich immer wieder auf den Boden der Tatsachen holst!

Zum Schluss geht noch ein sehr grosser Dank an meine Familie, welche mich stets unterstützt hat, jeden Schritt getragen und einige auch erst ermöglicht hat. Besonders herzlich danke ich dabei meinen Eltern Karla und Peter, meiner Schwester Sara und meinen Grosseltern Johanna und Max und Charlotte und Walter, für die immerwährende Hilfe, die Fürsorge, das Verständnis und die vielen, schönen, bereichernden und prägenden Momente.

MIX
Papier aus verantwortungsvollen Quellen
Paper from responsible sources
FSC® C105338

Printed by Books on Demand GmbH, Norderstedt / Germany